Xudong Xie

Assessment of an ultramicroelectrode array (UMEA) sensor for the determination of trace concentrations of heavy metals in water

Karlsruher Mineralogische und Geochemische Hefte

Schriftenreihe des Instituts für Mineralogie und Geochemie,

Universität Karlsruhe (TH)

Band 26

Assessment of an ultramicro-electrode array (UMEA) sensor for the determination of trace concentrations of heavy metals in water

von
Xudong Xie

universitätsverlag karlsruhe

Dissertation, Universität Karlsruhe (TH),
Fakultät für Bauingenieur-, Geo- und Umweltwissenschaften, 2004

Anschrift des Autors:
Xudong Xie
Institut für Mineralogie und Geochemie
Universität Karlsruhe (TH)
D-76128 Karlsruhe

Anschrift der Schriftleitung:
Karlsruher Mineralogische und Geochemische Hefte
Institut für Mineralogie und Geochemie
Universität Karlsruhe (TH)
D-76128 Karlsruhe

Impressum

Universitätsverlag Karlsruhe
c/o Universitätsbibliothek
Straße am Forum 2
D-76131 Karlsruhe

www.uvka.de

Print on Demand

ISSN 1618-2677
ISBN 3-937300-48-1

Abstract

Rapid development in silicon technology and microelectronics has nowadays enabled the mass fabrication of microelectrode arrays with well-defined and reproducible geometries on micron scales by using thin-film technology. Compared to conventional macroelectrodes, microelectrodes possess several attractive features, such as high mass transport, reduced charging current, immunity to ohmic drop, high signal-to-noise ratio, etc., which are very advantageous for developing decentralised analytical equipments in trace metal analysis.

A novel iridium-based ultramicroelectrode array chip (Ir-UMEAs) was recently designed and fabricated by means of microlithographic techniques by one of the partners (Fraunhofer Institute of Silicon Technology, ISIT, Germany) of this joint research project. The Ir-UMEA chip is made up of 4048 individual ultramicroelectrodes (working electrodes) arranged in four separate arrays. Each array consists of 46 x 22 = 1012 disc shaped microelectrodes with a diameter of 1.8 μm each, giving a total electrode area of 2575 μm^2. The electrodes are recessed by 0.2 μm and are spaced with an interelectrode distance of 25 μm.

This study is mainly aimed at (i) investigating the electrochemical behavior and the analytical performance of the Ir-UMEAs, (ii) assessing the analytical parameters of square-wave anodic stripping voltammetric measurements (SWASV) using the Ir-UMEAs, and (iii) estimating the potential of the Ir-UMEAs for the development of an analytical instrument to determine trace metal concentrations in natural waters. Surface analytical techniques (SEM, ESEM, and AFM) showed that the Ir-UMEA chips are of high quality, without evident defects in manufacture. Cyclic voltammetric and chronoamperometric experiments proved that the electrochemical behavior of the Ir-UMEAs is dominated by features characteristic for microelectrodes (sigmoidal voltammograms, steady-state current response, etc.). The quality of the Hg-film proved to be crucial in achieving a high analytical performance. Chronocoulometry was found to be the most suitable method in generating the Hg-film of good quality. An Ir-UMEA chip could be plated at least ten times without obvious loss in analytical performance, corresponding to approximately 500 measurements of trace metal concentrations in synthetic aqueous solutions. Using SWASV, detection limits of <0.1 µg/l could be obtained for the metals investigated (Cu, Pb, Cd, Zn). Precision and accuracy were found to be satisfactory (about ±10% of RSD). Though fouling of the arrays with time may become a problem, an ultrasonic bath proved to be very efficient in cleaning the surface of the electrodes in laboratory. Instrumental parameters, experimental conditions, and possible interferences were investigated and optimised. Interferences due to dissolved organic matter and/or intermetallic compounds were significantly reduced by means of medium exchange and standard addition. The results obtained with the Ir-UMEA sensor showed a reasonably good agreement with high resolution ICP-MS analyses, indicating that the combination of microelectrode arrays with SWASV has a great potential in developing a marketable field analytical system for the determination of heavy metal concentrations in natural waters.

However, for in situ and on-site measurements, the protection of the UMEAs from fouling (e.g., the integrating of the arrays in an anti-fouling membrane) seems to be inevitable. In order to avoid the shield effect caused by the overlapping of diffusion layers of the individual electrodes and thus to achieve a pure microelectrode behavior with the Ir-UMEAs, increasing the ratio of interelectrode distance to electrode diameter is recommended for further developments.

Acknowledgments

I would like to thank Prof. Dr. Doris Stüben for giving me the opportunity to make my PhD study at the Institute of Mineralogy and Geochemistry (IMG). Without her help the completion of this dissertation would not have been possible and my stay in Germany would not be so fruitful. I would also like to express my thanks to Prof. Dr. Rolf-Dieter Wilken for taking on the role of Korreferat.

Many thanks go to Dr. Zsolt Berner for his unselfish support during the course of this work. His valuable suggestion, critical review of the thesis, and kindly encouragement are very helpful.

Special thanks go to Dr. Stefan Norra for his help in using Geographic Information System (GIS, ArcView) and MapInfo programmes; Ms. Claudia Mößner for her effort in ICP-MS analysis; Mr. Peter Schaup for his various technical assistance; Dr. Gerhard Ott for solving computer problems; and Ms. Gesine Preuß for the preparation of laboratory work.

I am grateful to the following colleagues for their various helps: Prof. Heinz-Günter Stosch, PD. Dr. Thomas Neumann, PD. Dr. Jörg-Detlef Eckhardt, Andrea Stögbauer, Ms. Rosemarie Bender, Dr. Utz Kramar, Ms. Nytz Halina, Ms. Maria Tannhäuser, Hassan Aboughalma, and Michael Lauer etc..

In a word, I thank all the peoples who have provided me helps during the past several years. Because of their helps, I have spent a very special and nice time. Their hospitability will stay in my memory forever.

Finally, I wish to express my gratitude to my family: my parents, Yun Xie & Yuzhen Zang; my wife, Cuiling; and my son, Kaibo, who have given me a lot of love, encouragement, understandings, and energies.

Thanks to all of you

Table of Contents

List of Figures

List of Tables

List of Appendices

Abbreviations

AAS	Atomic absorption spectroscopy
ACSV	Adsorptive cathodic stripping voltammetry
AdSV	Adsorptive stripping voltammetry
AFM	Atomic force microscope
ASV	Anodic stripping voltammetry

CA	Chronoamperometry
CE	Counter electrode
CSV	Cathodic stripping voltammetry
CV	Cyclic voltammetry
DMG	Dimethylgloxime
DOC	Dissolved organic carbon
EDX	Energy dispersive X-ray analysis
ESEM	Environmental scanning electron microscope
HR-ICP-MS	High resolution inductively coupled plasma mass spectrometry
IC	Ion chromatography
ICP-MS	Inductively coupled plasma mass spectrometry
INA	Instrumental neutron activation
Ir/Pt-UMEA	Iridium-platinum ultramicroelctrode array
Ir-UMEA	Iridium ultramicroelectrode array
LSV	Linear sweep voltammetry
PECVD	Plasma-enhanced chemical vapour deposition
PSA	Potentiometric stripping analysis
Pt-UMEA	Platinum ultramicroelctrode array
RC	Resistance capacity
RE	Reference electrode
RSD	Relative standard deviation
SCE	Saturated calomel electrode
SD	Standard deviation
SEM	Scanning electron microscope
SW	Square wave technique
SWASV	Square-wave anodic stripping voltammetry
TXRF	Total reflection X-ray fluorescence
UME	Ultramicroelectrode
UMEAs	Ultramicroelectrode arrays
UMEs	Ultramicroelectrodes
WE	Working electrode

1. Introduction

There is a growing concern about the contamination of natural waters with heavy metals due to the detrimental impact of those elements on the environment. Considerable amounts of heavy metals are released into aquatic systems (lakes, rivers, oceans) from various anthropogenic sources, like municipal sewage effluents, garbage incineration, combustion of fossil fuels, metallurgical industry, mining, agricultural runoff, etc. (MOORE & RAMAMOORTHY 1984, ALLOWAY 1993). Even at low concentrations (10^{-7}-10^{-11} M) (STUMM 1996), heavy metals may present a severe hazard to the normal functioning of aquatic systems as some of them are very toxic, are not biodegradable, and are involved in biogeochemical cycles (FIEDLER & ROESLER 1993). In natural waters, heavy metals exist generally in particulate and dissolved form, such as inorganic species, organic complexes, and metal ions adsorbed onto a variety of colloidal particles (FLORENCE 1986, BUFFLE & HORVAI 2000). Any variation in the speciation of heavy metals will influence their bioavailability, toxicity, and geochemical behavior (e.g., mobility, adsorption, and precipitation) in natural waters (TESSIER & TURNER 1995, HILL 1997, MILLERO & PIERROT 2002).

For the assessment and a better understanding of the pollution of natural waters with heavy metals, reliable and robust analytical techniques with low detection limits are required. Various analytical approaches have been used for the determination of trace concentrations of heavy metals, such as atomic absorption spectroscopy (AAS), inductively coupled plasma mass spectrometry (ICP-MS), ion chromatography (IC), total reflection X-ray fluorescence (TXRF), instrumental neutron activation (INA), electrochemical stripping analysis (e.g., ASV, PSA), etc. (KOUNAVES 1990, PALCHETTI ET AL. 1999). Ever since, stripping analysis has been always regarded as a very powerful electrochemical technique for trace metal measurements. The strength of stripping voltammetry is in its extremely low detection limits (10^{-10}–10^{-12} M), its multi-element and speciation capabilities, and its suitability for on-site and in situ applications (KISSINGER & HEINEMAN 1996). Using this method, trace concentration of heavy metals could be directly determined without any pretreatment of samples, such as external preconcentration or matrix separation as commonly required in other analytical techniques (STATHAM ET AL. 1985, GRASSHOFF ET AL. 1999). Its extremely high sensitivity is attributed to its inherent ("built-in") preconcentration step (WANG 2000). For the determination of saline samples with high concentrations of matrix elements, like seawater, ICP-MS cannot be used without dilution or analyte extraction (BRENNER & TAYLOR 1992, ALLEN ET AL. 1997, GAO ET AL. 2002), although it provides very low detection limits. In contrast, for such samples, stripping analysis can be easily applied, the salinity of samples being even of advantage because it serves as natural electrolyte (NUERNBERG 1984, DANIELE ET AL. 1989, ACHTERBERG & BRAUNGARDT 1999).

One of the most important recent advances in stripping analysis has been its coupling with ultramicroelectrodes. This combination eliminates the need for forced convection during deposition step (WOJCIECHOWSKI & BALCERZAK 1991), minimizes variations due to natural convection (GAO ET AL. 1995), facilitates assays of very small sample volumes (WOJCIECHOWSKI & BALCERZAK 1990, WANG ET AL. 1999, EMONS ET AL. 1999), permits stripping analysis in poorly conductive media (WIGHTMAN 1988, CISZKOWSKA & OSTERYOUNG 1995) or speciation determination in low ionic-strength solutions (TERCIER-BAEBER & BUFFLE 1999), and increases the sensitivity of measurement (detection limit 10^{-12} M) (KISSINGER & HEINEMAN 1989). Using microelectrodes, it is quite possible to develop an in situ and on-site portable analytical system for the trace metal analysis, because the design and operation are greatly simplified in comparison with conventional stripping analysis by means of macroelectrodes, in accordance with the requirements of a decentralised assay (TAILLEFERT ET AL. 1999, WANG 2000, BUFFLE & HORVAI 2000). In situ and on-site measurement provides a lot of advantages, such as elimination of the artefacts due to sample handling (e.g., sampling, transportation, storage, and sample pre-treatment) (MART 1982), capacities to measure metal speciation, real-time monitoring, and low cost (BUFFLE ET AL. 1997).

In this study, the analytical performance of a newly developed iridium-based ultramicroelectrode array (UMEAs) sensor has been investigated for the determination of trace concentrations of heavy metals in water. The work consists of four main parts:

- Characterization of ultramicroelectrode arrays (UMEAs)

- Evaluation of the analytical performance of UMEAs in the determination of trace concentrations of heavy metals in aqueous solutions
- Optimisation of experimental parameters
- Application of UMEAs in the determination of heavy metals in natural waters

The UMEA chips, fabricated by modern microphotolithography technique were characterized using surface analysis (SEM, ESEM, and AFM) and electrochemical methods, like cyclic voltammetry (CV) and chronoamperometry (CA). The aims of these investigations were to check the quality of fabrication and to get information about the electrochemical behaviour of the UMEAs. After determining the parameters for the generation of an optimum mercury film on electrodes, the analytical performance of UMEAs in the determination of trace concentrations of heavy metals, such as sensitivity, reproducibility, accuracy, and lifetime were tested. In order to get successful measurements using the UMEA sensor, instrumental parameters, experimental conditions, and interferences were examined and optimised. As a novel analytical approach, the UMEA sensor was tested by determining the trace concentration of heavy metals in natural water samples and the results were compared with high resolution inductively coupling plasma mass spectrometry (HR-ICP-MS). The potential for development of a portable in situ and on-site analytical system was estimated.

2. Review of research

2.1 Development of microelectrodes

The importance of microelectrodes attracted the attention of electrochemists firstly at the beginning of 1980s (SWAN 1980, WIGHTMAN 1981). However, the advantageous properties of microelectrodes have been widely recognized only since the developments in microelectronic technologies made it possible to reliably measure very low currents and to construct reproducibly very small electrodes (WIGHTMANN 1988). The exact definition of microelectrodes is ambiguous (AOKI 1993) and it is very difficult to give a definition in terms of precise limits of its characteristic dimensions. Nevertheless, electrodes with dimensions of tens of micrometers or less, down to submicrometer are often called microelectrodes (STULIK ET AL. 2000). Compared to conventionally sized electrodes (1-5 mm radius) (BOND 1994) and macroelectrodes (radius r ≥ 100 µm) (BUFFLE & HORVAI 2000), microelectrodes possess a lot of attractive features, such as high mass transport, immunity to ohmic drop, reduced changing current, and small RC constant (PONS & FLEISCHMANN 1987, AOKI 1993, FLETSCHER 1994, WIGHTMAN 1989, BOND 1994). As a result, microelectrodes have opened new possibilities in the research fields of electrochemistry, biotechnology, medicine, and environmental science (CAMMANN ET AL. 1991, HEINZE 1993, JANATA 1992, RYAN ET AL. 1994, WOLLENBERGER ET AL. 1995, ANDERSON & BOWDEN 1996). In the last two decades, several studies were published dealing with the theory, fabrication, and application of microelectrodes. Microelectrodes, including single- and composite microelectrodes, were produced in diverse geometries, with different electrode materials, and using various fabrication techniques. Due to the very low current obtained with single microelectrodes (typically in pA to nA range) (FORSTER 1994) and the difficulty in construction of high sensitive instrumentation (FLETCHER & HORNE 1999), composite microelectrodes (array and ensemble) (TALLMAN & PETERSEN 1990) have been widely used.

The theory of electrochemical behavior of single microelectrodes is now well established, with even precise analytical expressions (WIGHTMAN 1989, FLEISCHMANN ET AL. 1987). However, the theory of composite microelectrodes is still not developed due to its complexity, especially for ensemble microelectrodes. To some extent, this makes it difficult to explain some experiment results obtained with microelectrode arrays. Several models have been established to simulate the mass transport behavior of array electrodes. The chronoamperometric response of microdisk arrays was firstly estimated using analytical expressions under simplifying assumptions (LINDEMANN & LANDSBERG 1971, GUESHI 1978). Later, a digital simulation approach was proposed using the explicit finite-difference technique (RELLER ET AL. 1982). However, this model was suitable only for short and long timescales; for intermediate time the currents obtained were inaccurate. Using the hopscotch algorithm, a digital simulation of the chronoamperometric response with an empirical analytical formula was presented with the accuracy better than 2% for hexagonal microdisc arrays with inlaid geometry (SHOUP & SZABO 1984). This model has been often applied to check the amperometric current response of UMEA at intermediate timescales (SEDDON ET AL. 1994, KOUNAVES ET AL. 1994, TERCIER-WAEBER ET AL. 1999). An analytical expression was reported for both ordered (square and hexagonal) and random arrays (SCHARIFKER 1988). Recently, a 3D simulation of the chronoamperometric response based on finite element resolution for a regular microelectrode array (BERIET ET AL. 2000), a simulation of cyclic voltammetric response at a regular microdisc electrode array, (LEE ET AL. 2001), and a simulation model concerning with the effect of the recess depth of microelectrodes on the current response (CASSIDY ET AL. 1985, BOND ET AL. 1988, FERRIGNO ET AL. 1997, BARTLETT ET AL. 1998) were reported, respectively. The fundamental features of mass transport to array microelectrodes, as well as the conditions and timescales whereby individual electrodes in an array act independently from each other were described by SCHARIFKER (1991).

Unlike single microelectrodes, for microelectrode arrays, the geometry of microelectrodes, i.e. the ratio (w/d) of interelectrode distance or spacing (w) to the electrode dimension (e.g., electrode diameter, d) have to be considered, in order to get a collective current response while maintaining the advantageous features of single microelectrodes. There are several articles which discuss theoretically the amperometric current response and packing density of microelectrode arrays with different geometries. It is suggested that arrays with a low packing density (d » 2r, r, electrode radius) yield the multiple current response of single microelectrodes, while closely packed arrays ($d \approx 2r$) resemble the behavior of macroelectrodes of the same total area (WEBER 1989, MORF 1996, MORF & ROOIJ 1997, MORF 1997). There is no consensus about the optimal geometry of composite electrodes. It is reported that the interelectrode spacing should be at least six electrode diameters ($w/d \geq 6$) in order to avoid the overlapping of the diffusion layers of adjacent microelectrodes (CAUDILL ET AL. 1982, WEST 1993, NASCHIMENTO ET AL. 1997, BARTLETT & TAYLOR 1998). The study of SANDISON ET AL. (2002) confirmed that, for microelectrode arrays with the same w/d, the smaller the individual microelectrode is the greater is the linear component of the mixed diffusion layer. A minimum interelectrode spacing of approximately $w/d = 20$ is required to ensure fully radial diffusion to microelectrode arrays with an individual electrode diameter of 5 µm (LEE ET AL. 2001). For determination of heavy metals in water, Ir-UMEAs were fabricated in microlithographic technique with w/d ratios of 10 and 30 (d = 5 µm, w = 50, 150 µm, separately). The shield effect was found to occur at a w/d ratio of 10 with these chips (BELMONT ET AL. 1996). For screen-printed carbon microelectrode array with w/d ratios of 6.7, 13.3, and 20.0 (d = 15 µm, w = 100, 200, 300 µm, respectively), the cyclic voltammetric and amperometric experiments showed that the shield effect occurred at a w/d ratio of 6.7 (SEDDON 1994). Arrays of gold minigrids with ratios of spacing to width varying from 1.5 to 14.4 revealed a voltammetric behaviour which ranged from quasi steady state to steady state, suggesting that the ratio of spacing to width should be equal to or greater than 14, in order to maintain the behaviour of microelectrodes (WU 1993). The study of FLETSCHER & HORNE (1999) about random assemblies of microelectrodes (RAM electrodes, d = 7 µm) demonstrated that a ratio of distance to radius greater than 20 ($d/r \geq 20$) is needed for negligible diffusive interferences. Some more recently developed microelectrode arrays with various geometries are also described below.

2.2 Applications of microelectrodes in the determination of trace metals

Coupled with stripping analytical methods and computer-controlled miniaturized instrumentation, microelectrodes show a great potential in the development of portable analytical instruments for the determination of trace metals in natural waters (BRETT 1999, WANG 1999). Thanks to the advantageous features of microelectrodes, microelectrodes are suitable for in situ and on-site measurement of trace concentration of heavy metals and difficulties encountered in the stripping analysis with macroelectrodes can be overcome (VITRE ET AL. 1991). The high mass transport accrued by radial diffusion of microelectrodes makes the natural convection (BELMONT-HEBER ET AL. 1998, ADELSALAM ET AL. 2002) less problematic and thus stirring becomes unnecessary during the preconcentration step (WOJCIECHOWSKI & BALCERZAK 1990). The small double layer capacitance of microelectrodes resulting from their small electrode surface area significantly reduces the charging current. The ratio of faradaic to non-faradaic current is greater and the detection limit is enhanced (WEHMEYER ET AL. 1985). The small currents generated by microelectrodes result in a negligible ohmic drop which allows stripping analysis to be performed also in high resistive media (HOWELL & WIGHTMAN 1984, DANIELE ET AL. 1989, WONG & EWING 1990, CISZKOWSKA & STOJEK 1999) or even without the need for supporting electrolyte, minimizing the chemical contamination. Square wave pulse technique is commonly applied in stripping analysis due to its high speed and enhanced sensitivity of measurement (CHRISTIE ET AL. 1977, OSTERYOUNG & OSTERYOUNG 1985, KOUNAVES ET AL. 1987, WECHTER & OSTERYOUNG 1989BRETT 1999). The use of square wave technique makes it possible to carry out the trace metal analysis in the presence of oxygen (KOUNAVES ET AL. 1987, WOJCIECHOWSKI & BALCERZAK 1990, KOUNAVES & DENG 1993, WANG ET AL. 1993, BRETT ET AL. 1994, PALCHETTI ET AL. 1999). As a result, the deaeration step which has to be used for conventional macroelectrodes can be obviated.

Lately, diverse sensors have been designed for the trace metal analysis, using either microlithographically fabricated or screen-printed microelectrode arrays. The proposed analytical techniques are chiefly based on the principles of stripping voltammetry (anodic-, cathodic-, and adsorptive voltammetry) and stripping potentiometry (PSA) (OSTAPCZUK 1993).

2.2.1 Voltammetric sensor

There are several papers which report on the development of mercury-plated iridium-based microelctrode array sensors for the determination of heavy metals using square wave anodic stripping voltammetry (SWASV). BELMONT ET AL. (1996) reported a photolithographically produced iridium-based microelectrode array which integrates 100 microelectrodes with a diameter of 5 µm each, and an interelectrode distance of 150 µm. The arrays were characterized by cyclic voltammetry, chronoamperometry (CV), scanning electron microscope (SEM), and atomic force microscope (AFM), showing a good fabrication quality. A microelectrode array chip could be used for the operations of deposition/dissolution of mercury film at least 10 times. For a preconcentration time of 15 min, a detection limit of 50 pM was achieved for trace metal measurement in synthetic solution. Another Ir based microelectrode array by FEENEY ET AL. (1998) consists of 25 microdiscs with a diameter of 10 µm each and an interelectrode distance of 100 µm. It was tested for the determination of Cd in aqueous solutions. Satisfactory analytical results were obtained at concentration levels of 20 to 100 µg/l Cd. Atomic force microscope (AFM) analysis showed that the surface of microelectrode array was covered by dirty accumulations after several measurements, which proved to be mercurous chloride (Hg_2Cl_2) (JAGNER ET AL. 1996, NOLAN & KOUNAVES 1998, NOLAN & KANAVES 1998, NOLAN & KOUNAVES 1999, NOLAN & KOUNAVES 2000). An Ir-UMEA chip, consisting of 30 x 30 microdiscs with a diameter of 5 µm each and a spacing of 180 µm each other was described by SILVA ET AL. (1999). Nearly identical cyclic voltammograms were obtained at scan rates of 5 and 500 mV/s, indicating that the interelectrode distance of 180 µm is sufficient large. The arrays could sustain the operation of mercury film plating/deplating at least 15 times and a detection limit of 0.5 µg/l could be achieved for the determination of Cd and Pb in synthetic aqueous solutions. A newly formed mercury film was stable in the measurement of 1 µg/l Cd in 0.01 M KNO_3 solution for about 5 hours.

In addition to the commonly measured heavy metals (Cu, Pb, Cd, and Zn), there are also reports on the determination of other elements like As, Ni, Se, Hg, etc., using microelectrode array. For the on-site analysis of arsenic (As^{3+}) in ground water, instead of Hg-plated Ir-based microelectrode array, a nafion-coated mercury free Au-based microelectrode array including 564 microelectrodes with a diameter of 12 µm and a center-to-center spacing of 58 µm was used (FEENEY & KOUNAVES 2000, FEENEY & KOUNAVES 2002). As^{3+} in aqueous solutions was analysed after the deposition of an As^0 monolayer by SWASV. A detection limit of 0.05 µg/l As and precision of 2.5 % (RSD) could be achieved. For the direct determination of Cu and Hg, a mercury free Ir-based microelectrode array was reported by Nolan & Kounaves (1999). The determination of selenium (IV) using Au-based microelectrode array was documented by Tan & Kounaves (1998).

2.2.2 Potentiometric sensor

Potentiometric stripping analysis (PSA) is a good alternative to trace metal analysis (JAGNER 1983, HANSEN 1991, JAGNER ET AL. 1993) because some of the difficulties encountered in voltammetric stripping analysis, such as those due to the presence of organic matter and electroactive compounds, are less problematic (TERCIER & BUFFLE 1990, OSTAPCZUK 1993, DANIELE ET AL. 2000). With a screen-printed carbon microelectrode array (EDINBURGH 1992) containing four carbon microdiscs with a diameter of 15 µm each and an average interelectrode distance of 160 µm, trace metals were determined using both square wave voltammetry (SWASV) and potential stripping analysis (PSA) (WANG 1993). The array revealed non-linear diffusion behavior in cyclic voltammetric study. The chip showed good analytical performance at concentration levels of above 25 µg/l by placing a 250µl sample droplet on the array strip without deaeration and stirring. With the same type of carbon microelectrode array, a hand-held Pb analyser was designed and constructed which allows the measurement of Pb concentrations of 20 - 300 µg/l by potentiometric stripping analysis (YARNIZKY 2000). Using adsorptive potentiometric stripping analysis (KALVODA 1994, ZUHRI & VOELTER 1998, BRAININA ET AL. 2000), an Ir-based microelectrode array with 564 individual microelectrodes (10 µm in diameter, 68 µm in interelectrode distance) per array was successfully employed for the determination of trace concentrations of nickel in the presence of dimethylgloxime complexing agent (DMG). Trace concentrations of Ni could be measured by accumulating the Ni-DMG complex on the electrode surface under an adsorption potential of -0.2 V and a time frame of 5 - 30 s. The stable PSA signals and low detection limit (0.5 µg/l with 30 s adsorption time) indicate the potential of microelectrode array in determining trace metals which cannot be electrolytically plated (WANG 1999).

2.2.3 In situ sensor

In situ measurement based on stripping analysis is a very promising approach for the speciation determination of heavy metals in natural waters, because the artefacts resulted from sample handling could be significantly minimized (FLORENCE 1986, BUFFLE 1993, BUFFLE 1997, BUFFLE & HORVAI 2000). Recently, several reports were published which concerned with the development of in situ sensors using microelectrodes. A novel gel-integrated Ir-based microelectrode array sensor was developed for in situ measurements of trace metals (Pb, Cd, and Mn) in lake water (BELMONT ET AL. 1998, TERCIER-WAEBER ET AL. 1998). The device consists of a grid of 10 x 10 Ir microdisc shaped microelectrodes with a diameter of 5 µm each and a center-to-center spacing of 150 µm, which is coated with a 300-600 µm thick 1.5% agarose gel. Because the gel is inert to heavy metals, it was used as a dialysis membrane, allowing diffusion of metal ions and small metal complexes, while hindering colloids and macromolecules to the surface of the array (TERCIER & BUFFLE 1996). Consequently, the array is protected from fouling caused mainly by suspended matters and organic compounds (HOYER ET AL. 1987, TERCIER & BUFFLE 1993, TERCIER ET AL. 1998). The optimum conditions and characteristics of the gel-coated array were studied by recording voltammetric signals under different conditions, such as convection (AMATORE & FOSSET 1992), temperature, pressure, DOC content, etc.. The good reproducibility of the results and the low detection limit (50 pM) demonstrate the suitability of the sensor for in situ measurement of trace metals in natural waters (LAFLEUR ET AL. 1990, BOND 1994, TERCIER-WAEBER & BUFFLE 1996, TERCIER-WAEBER & BUFFLE 1999, TERCIER-WEBER & BUFFLE 2000, DANIELE ET AL. 2002, HOWELL ET AL. 2003). A drawback of the membrane is that metal species diffusing into the gel-layer are not well-defined (dissolved, colloidal, particulate, etc.), and in situ measurements need to be compared to total analyses of filtered and unfiltered waters to determine the metal speciation measured by voltammetry (Tercier et al. 1995, Belmont et al. 1998).

Another Ir-based microelectrode array sensor for in situ screening of heavy metals (Cu, Pb, Cd, and Zn) in ground water was reported by HERDAN et al. (1998). The array was manufactured microlithographically, consisting of 20 microdisc shape microelectrodes (10 µm in diameter and 100 µm in spacing). An external miniature solid-state Ag/AgCl reference electrode was used in the three-electrode configuration to eliminate leakage of electrolyte (Cl^-) (NOLAN & KOUNAVES 1998, NOLAN & KOUNAVES 1999, NOLAN & KOUNAVES 2000). Without any pre-treatment of the samples, a reasonable representation of heavy metal concentrations for initial, rapid, and order-of-magnitude, field screening of heavy metals was provided by the sensor. Measurements were validated by means of the ASV measurement with laboratory-based glassy carbon macroelectrode (3 mm in diameter) and ICP-MS analysis. The inconsistency results were explained to be due to (i) the fouling of electrodes caused by adsorption of large inorganic colloids and suspended matter onto the electrode surface (FLORENCE 1986), (ii) the influence of organic surfactants or complexing agents, (iii) the interference of intermetallic compounds, such as Cu/Zn compounds (BRETT ET AL. 1996), and (iv) the saturation of Hg-film of microelectrodes by the high concentrations of heavy metals in the water samples.

Remote electrochemical sensors for the measurements of trace metals (Cu, Pb, Hg, Se) in natural waters were reported by WANG ET AL. (1995, 1998) and DANIELE ET AL. (2000). These analytical devices were based on single gold fiber microelectrodes (diameter, 100 µm) and stripping potentiometry. The studies have shown that the advantages of using microelectrodes in stripping analysis, such as no necessity of deoxygenation, of forced convection, and of electrolyte addition are preserved also by single electrode devices with a relative large dimension (100 µm). This special-designed mercury-free sensor was applied in the determination of trace metals in both seawater and ground water with low detection limits.

All the microelectrodes reported above have diameters larger than 5 µm, i.e., relative large dimensions. Until now no references were reported on the development of heavy metal sensors using microelectrodes smaller than 2 µm in diameter, though, theoretically, the advantageous features of microelectrodes becomes more characteristic, if the dimension of microelectrodes are reduced (e.g., the rate of mass transport increases as the electrodes size decreases) (WIGHTMAN 1981, PONDS & FLEISCHMANN 1987, TERCIER & BUFFLE 1993, BUFFLE & HORVAI 2000). This deficiency may be probably due to the technical difficulty in producing microelectrode array of very small size and of good quality. In this study, a novel Ir-UMEA was fabricated microlithographically using the ion-milling technique. The chip consists of 4048 UMEs with a diameter of 1.8 µm each and an interelectrode distance of 25 µm disposed in four separate arrays (XIE ET AL. 2004). In such an array design, one or more elements of arrays can be used at a given time of measurements, while the remainder can be used at a later time, increasing thus the lifetime of arrays. In the following a systematic study of the Ir-UMEAs is presented in the respects of fabrication process, electrochemical characteristics, analytical performance, and application potential in natural waters.

3. Theory

3.1 Electrochemical methods

3.1.1 Electrochemical cell

Electrochemical experiments are usually performed in an electrochemical cell, which consists mainly of the three electrodes, i.e. the working-, reference-, and counter electrode (Fig. 1). The working electrode (WE) is the electrode at which the reaction of interest occurs. In stripping analysis of heavy metals, mercury is still mostly used as working electrode material, because it has a high hydrogen overvoltage that greatly extends the cathodic potential window and possesses a highly reproducible, readily renewable, and smooth surface (HENZE & NEEB 1986). The reference electrode (RE) provides a known and stable potential, against which the potential of working electrode is compared. The most common reference electrode systems used in aqueous solutions are the silver-silver chloride (Ag/AgCl) and the saturated calomel (Hg/Hg_2Cl_2) electrode, which have an electrode potential independent of the composition of electrolyte (KELLNER ET AL. 1998). The counter electrode (CE), also called auxiliary electrode, is a current-carrying electrode, via which the current is

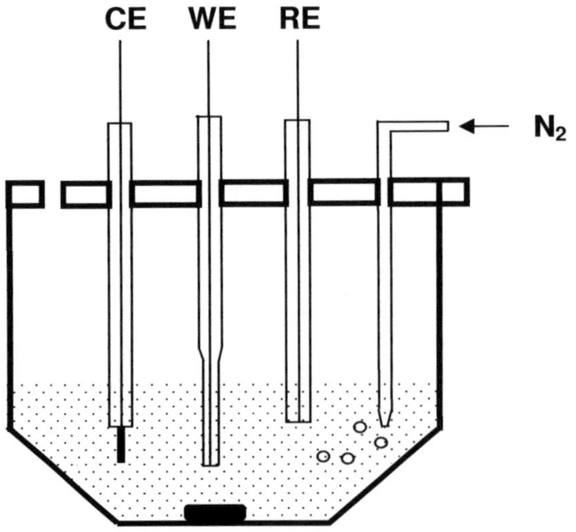

Fig. 1: Electrochemical cell with the working-, reference-, and counter electrode

measured. Inert conducting materials, such as a platinum wire or a graphite rod are often used as counter electrodes (WANG 1995). However, for microelectrodes, a micro flow through cell and a two-electrode configuration can be used due to the immunity to ohmic drop (FITCH & EVANS 1986).

3.1.2 Stripping analysis

Thanks to the inherent preconcentration step, stripping analysis is an extremely sensitive electrochemical technique, making it suitable for the determination of trace elements. In general, stripping analysis includes anodic stripping voltammetry (ASV), cathodic stripping voltammetry (CSV), adsorptive stripping voltammetry (AdSV), and potentiometric stripping analysis (PSA) (WANG 2000). Using stripping analysis more than 30 elements can be successfully measured (Fig. 2) (TERCIER & BUFFLE 1993, SAMUEL & KOUNAVES 1990, TESSIER & TURNER 1995).

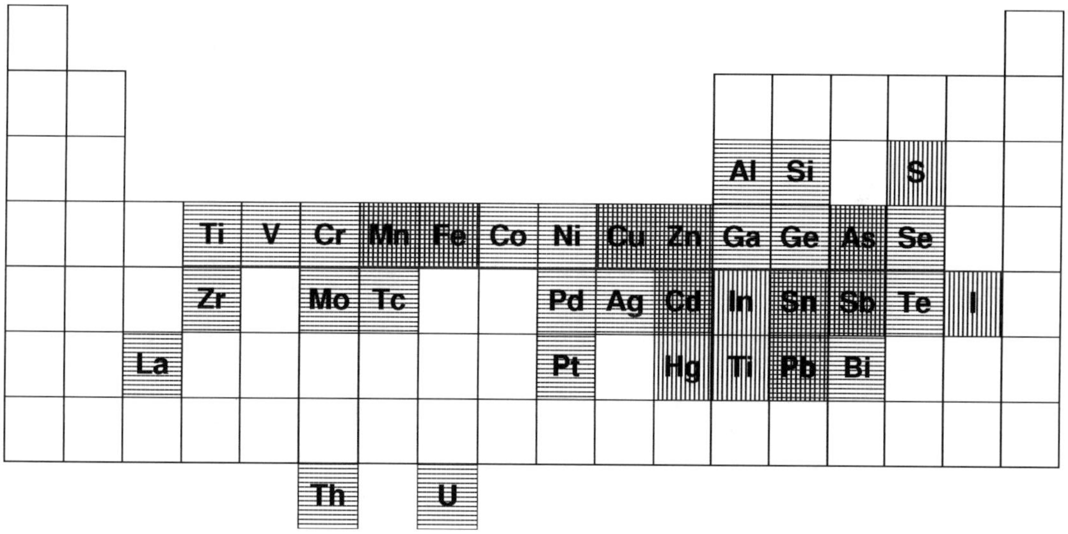

Fig. 2: Elements measured by anodic stripping voltammetry (ASV) and adsorptive cathodic stripping voltammetry (ACSV): ▤, ASV; ▥ , ACSV; ▦ , ASV or ACSV (reproduced from TESSIER & TURNER 1995)

3.1.2.1 Anodic stripping voltammetry (ASV)

The most widely used technique is the anodic stripping voltammetry (ASV) (Fig. 3). The application of ASV is limited to metals which can be reduced to the metallic state at potentials within the stability boundaries of water and mercury (between 0 and – 1.5 V), and which at the same time are soluble in mercury. A group of some 15 metals meets the requirements (VAN DEN BERG 1988). In the first step, metal ions in samples are preconcentrated on the working electrode as amalgam by cathodic deposition at a controlled potential and a given time.

$M^{n+} + ne^{-} + Hg \rightarrow M(Hg)$ (amalgam)

Following this preconcentration step, the amalgamated metals are reoxidized, and are stripped out of the electrodes by scanning anodically the potential.

$M(Hg) \rightarrow M^{n+} + ne^{-} + Hg$ (redissolution)

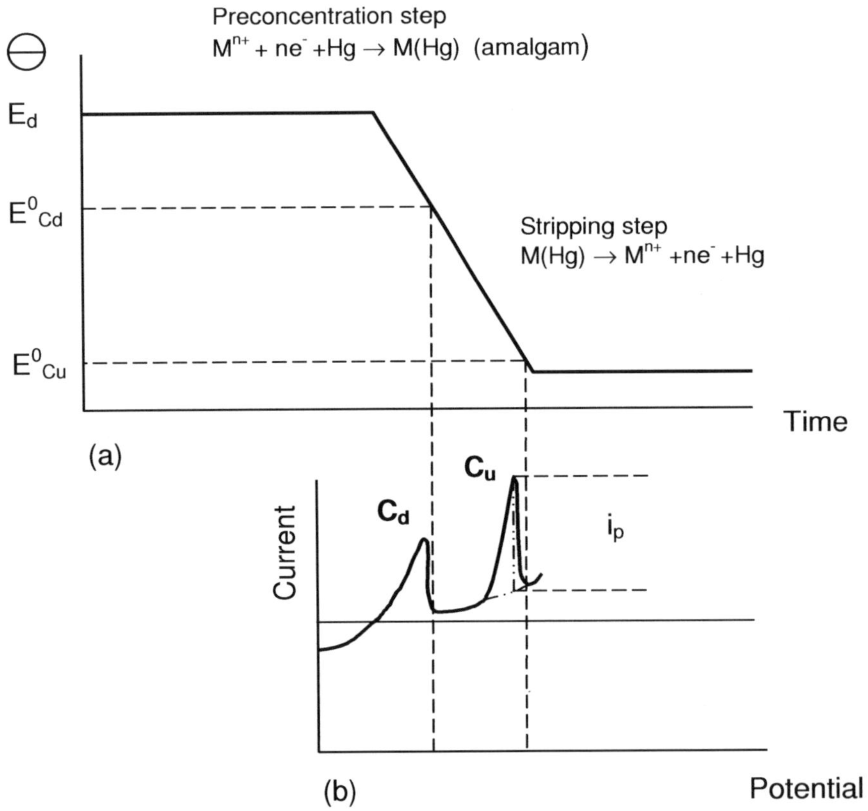

Fig. 3: Principle of anodic stripping voltammetry (ASV): (a) the potential–time waveform, (b) the resulting voltammogram. E_d, deposition potential; E^0_{cd}, standard potential of Cd; E^0_{cu}, standard potential of Cu; I_p, peak current (reproduced from WANG 2000)

From the peak potentials and the peak currents of the recorded voltammograms, the trace metals can be measured qualitatively and quantitatively. It should be noted that stirring (forced convection) and deoxygenation, which have to be used for macroelectrodes become less important for microelectrodes, particularly by using the SWASV approach.

3.1.2.2 Potentiometric stripping analysis (PSA)

Potentiometric stripping analysis (PSA) is another very important electrochemical approach for trace metal analysis (JAGNER 1982, HANSEN 1991). Similar to anodic stripping voltammetry (ASV), it consists also of two steps: potentiostatic preconcentration and stripping. It differs from ASV mainly in the methods used for stripping (JAGNER 1983). After preconcentration of metal ions on the working electrode, the potentiostatic control is disconnected (no current passes through the electrode) and the amalgamated metals are reoxidized by chemical oxidants, such as Hg^{2+} and O_2, or by using a constant current. A potential-time stripping curve is recorded (Fig. 4a). A peak shaped curve (dt/dE vs. E) like that in ASV can be achieved using modern computer-controlled PSA instrumentation (Fig. 4b).

$M^{n+} + ne^- + Hg \rightarrow M(Hg)$ (amalgam)

$M(Hg)$ + oxidant (or constant current) $\rightarrow M^{n+}$ (reoxidation)

Compared to ASV, potential stripping analysis has several advantages, such as less sensitive to organic compounds, no necessity of deoxygenation, possibility of performing measurement with a very small volume of sample (KOMORSKY-LOVRIC & BRANICA 1993, GIL & OSTAPCZUK 1994, SOARES & VASCONCELOS 1995, RISO ET AL. 1997). These features are very advantageous for the development of a portable heavy metal analyser for field applications or for measurements of samples with complex matrix, for instance, biological samples (OSTAPCZUK 1993, JAGNER ET AL. 1993).

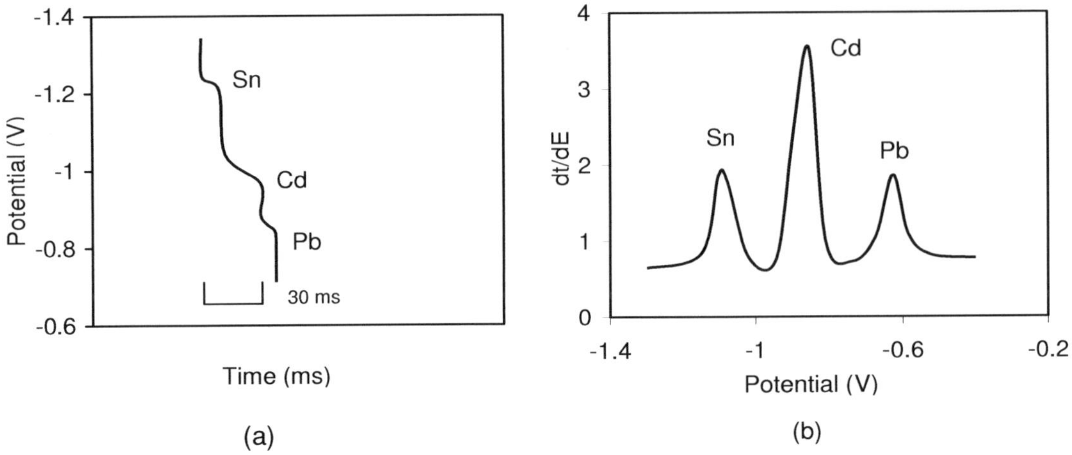

Fig. 4: The principle of poteniometric stripping analysis (PSA), (a) original potential-time stripping curve; (b) differential curve of dt/dE versus potential (reproduced from JAGNER 1983)

3.1.3 Cyclic voltammetry (CV)

Cyclic voltammetry (CV) is a potential-controlled reversal electrochemical experiment (CAMMANN 2001). It is performed by imposing a cyclic linear potential sweep, i.e., a triangular potential waveform on a working electrode in an unstirred solution (Fig. 5a). The resulting plot of current versus potential is termed cyclic voltammogram (Fig. 5b).

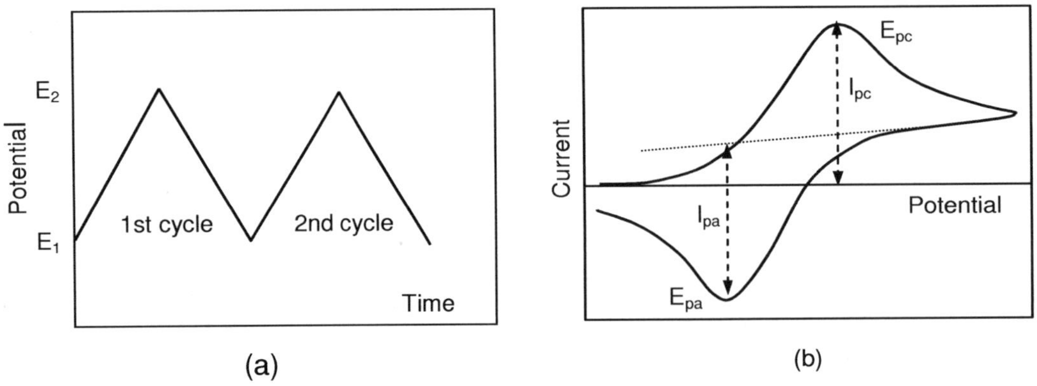

Fig. 5: Cyclic voltammetric experiment: (a) triangular potential waveforms; (b) cyclic voltammogram of a reversible reaction. I_{pc}, the cathodic peak current; I_{pa}, anodic peak current; E_{pc}, the cathodic peak potential; E_{pa}, the anodic peak potential

Cyclic voltammetry is mainly used for studying dynamic electrochemistry, such as thermodynamics and kinetics of electron transfer (HEINZE 1984, GOSSER 1993). For a reversible reaction system and a macroelectrode, the following information can be obtained from a cyclic voltammogram (Tab. 1) (MONK 2001).

Tab. 1 Cyclic voltammetric behavior of a macroelectrode for a reversible reaction

(i) The cathodic peak current (I_{pc}) is equal to the anodic peak current (I_{pa}), i.e., $I_{pc} = I_{pa}$
(ii) The peak potentials (E_{pc} and E_{pa}), are independent of the scan rate (ν)
(iii) The formal potential (E^0) is centred between E_{pc} and E_{pa}, $E^0 = (E_{pc} + E_{pa})/2$
(iv) The peak current (I_p) is proportional to the square root of scan rate ($\nu^{1/2}$)
(v) The separation between the peak potentials is 59 mV/n for an n-electron couple

The dependence of peak current on the scan rate is given by the Randles-Sevcik equation (BARD & FAULKNER 2001):

$$I_p = 0.4463nFA\left(\frac{nF}{RT}\right)^{1/2} D^{1/2}\, C\, \nu^{1/2}$$

where: n = number of electrons transferred

F = Faraday's constant (96,500 C mol^{-1})

A = electrode area (cm^2)

C = concentration (mol cm^{-3})

D = diffusion coefficient ($cm^2\ s^{-1}$)

R = the universal gas constant (8.314 J K^{-1} mol^{-1})

T = temperature (Kelvin)

For microelectrodes, a sigmoidal cyclic voltammogram is observed at a certain timescale, because the mass transport is dominated by radial diffusion. In steady state, a constant current (limiting current) is yielded, which is independent of scan rates.

3.1.4 Chronoamperometry (CA)

In chronoammperometric experiment, current is measured as a function of time. It involves stepping the potential of the working electrode from an initial value E_1 where no electrode reaction occurs, to a potential E_2 where the electrode reaction is complete, i.e. the concentration of electroactive species on the electrode surface is zero (Fig. 6a). For a planar electrode in unstirred solution, if the form of mass transport is only diffusion, then the current-time dependence is given by the Cottrell equation (WANG 2000):

$$I\,(t) = \frac{nFACD^{1/2}}{\pi^{1/2} t^{1/2}} = k\,t^{-1/2}$$

The current is proportional to the reciprocal of the square root of time (Fig. 6b). In contrast, for microelectrodes a time-independent current response is obtained after a short time of measurement due to the attainment of the steady state.

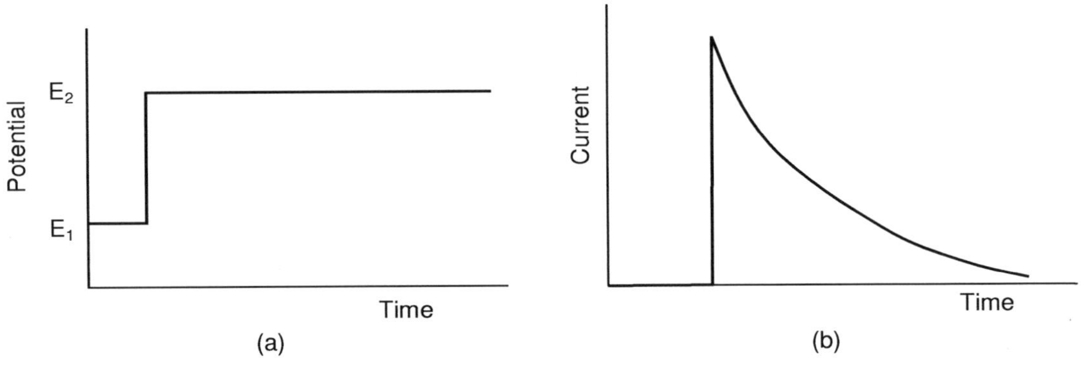

Fig. 6: Chronoamperometric experiment: (a) potential-time waveform, E_1, initial potential, E_2, final potential; (b) the resulting current-time response

3.1.5 Square wave technique

In voltammetry, when a potential is applied to the working electrode, current flows as a result of faradaic reaction and the charging of double layer capacitance of the electrode, including the faradaic current and the non-faradiac current (or the charging current). The faradaic current results from an oxidation or reduction of electroactive species, while the charging current does not involve any chemical reaction and is merely interference for the measurements. In order to reduce the detection limits, the ratio of faradaic current to charging current has to be increased. For this purpose, pulse voltammetric techniques, such as normal pulse-, differential pulse-, and square wave technique are commonly employed. However, the square wave technique affords more advantages, like high sensitivity, high speed, and no necessity of deoxygenation (OSTERYOUNG 1985, WOJCIECHOWSKI & BACERZAK 1990). As illustrated in Figure 7 the technique involves superimposing a symmetrical square wave on a staircase potential. By sampling the current twice during each square wave cycle, once at the end of the forward pulse (t_1) and once at the end of the reverse pulse (t_2), an enhanced net current is obtained (BUCHBERGER 1998, ALFASSI 1994). Compared to faradiac current, charging current decays exponentially with the time, which is

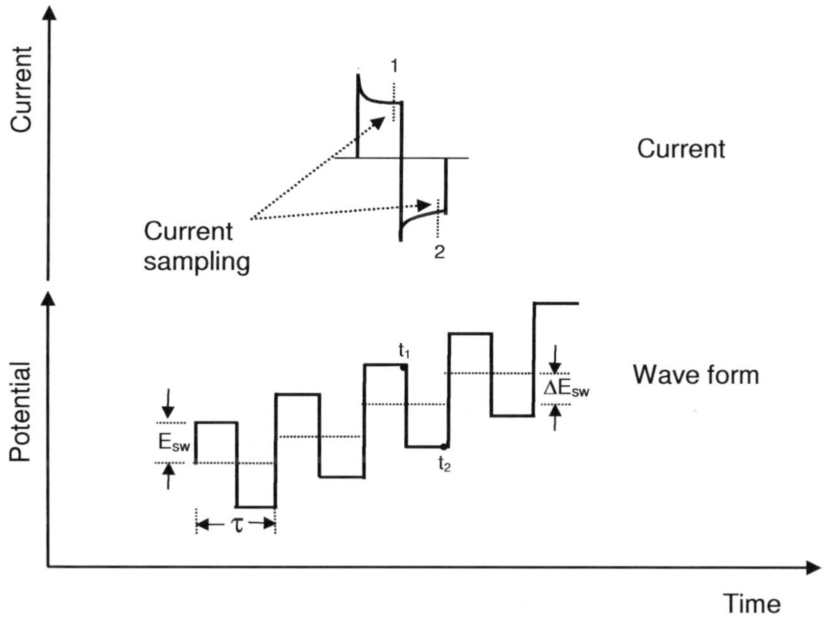

Fig. 7: Schematic representation of square wave pulse technique. E_{sw}, the amplitude; ΔE_s, step potential; τ, waveform period; and current measurement times, t_1 and t_2 (modified after ALFASSI 1994)

much quicker, especially for microelectrodes. In this way, a large faradaic current can be isolated from the charging current of the electrodes, increasing thus the sensitivity of measurements.

3.2 Theory of microelectrodes

3.2.1 Definition of microelectrodes

Microelectrodes are commonly known also as ultramicroelectrodes. There is no consistent opinion about its definition. Different conceptions are given here for reference:

- The term of microelectrode is reserved for electrodes with at least one dimension not greater than 25 µm (WANG 2000).
- It has become practice to use the term UME for disc electrodes with a diameter smaller than or equal to 20 µm (HEINZE 1993).
- Microdisc electrodes are electrodes with a radius not greater than 25 µm (BOND 1994).
- Electrodes whose characteristic length is less than 20 µm are often called ultramicroelectrode (AOKI 1993).
- Microelectrodes may be defined as electrodes whose critical dimension is in the micrometer range (FORSTER 1994).
- Microelectrode is an electrode with at least one dimension small enough that its properties, e.g., mass transport regime, are a function of size. Practically the critical dimension will generally fall in the range 0.1 to 50 µm (MONTENEGRO ET AL. 1991).
- Microelectrode is an electrode in which transport is controlled by diffusion only and for which $R/\delta \ll 1$ with R = electrode radius and δ = diffusion layer thickness (typical size is less than a few micrometers) (BUFFLE & HORVAI 2000).
- Microelectrode is any electrode whose characteristic dimension is, under the given experimental conditions, comparable to or smaller than the diffusion layer thickness. Under these conditions, a steady state or a pseudo steady state is attained. This is an operational definition of microelectrode (STULIK ET AL. 2000).

3.2.2 Microelectrode types

In general, microelectrodes can be classified as single microelectrodes and composite microelectrodes (Fig. 8). For single microelectrodes, there are different electrode types, such as disc, cylinder, band, ring, sphere, hemisphere, etc. (PONS & FLEISCHMANN 1987, MONTENEGRO ET AL. 1991, STOJEK 1991). Most of them are commonly fabricated by sealing fibers, foils into a non-conducting electrode body such as glass or epoxy holder (FORSTER 1993). As for composite microelectrodes, it can be generally divided into array electrodes and ensemble electrodes, depending on whether the surface of composite electrodes consists of uniform (array) or random (ensemble) dispersions of a conductor region within a continuous insulating matrix (TALLMAN 1990). The array electrodes are normally fabricated using microphotolithography (FIACCABRINO & KOUDELKA-HEP 1997, ANDREWS & HARRIS 1998, SUZUKI 2000, DATTA & LANDOLT 2000) and screen-printing (SEDDON ET AL. 1994, HART & WRING 1997), while the ensemble electrodes are produced by various

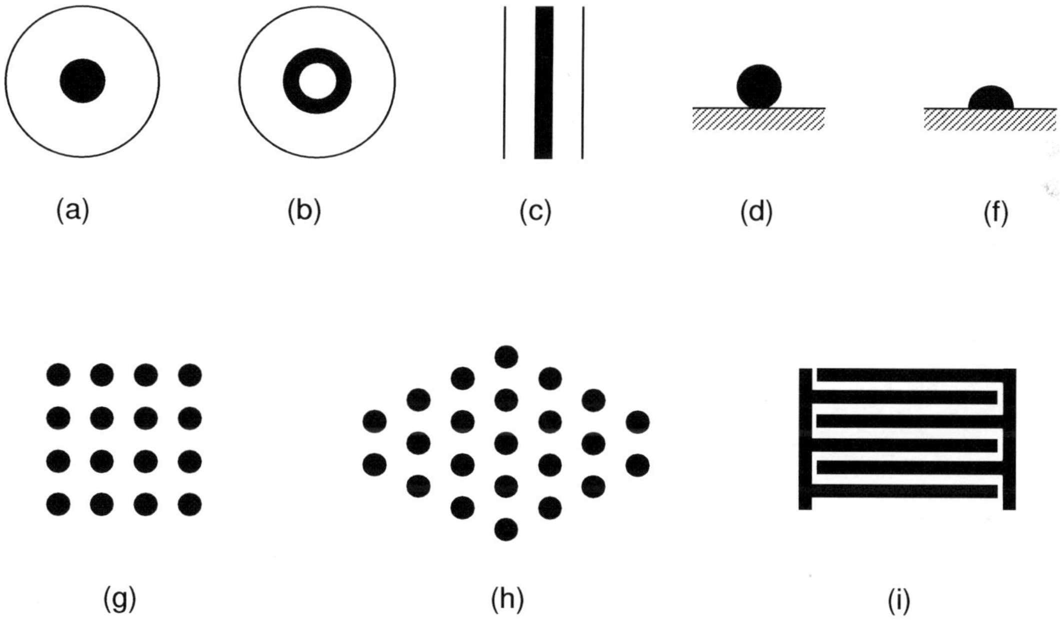

Fig. 8: The common types of microelectrodes (a), disc; (b), ring; (c), band; (d), sphere; (f) hemisphere; (g), square array; (h), hexagonal array; (i), interdigitated microband.

methods, such as by mixing or pressing a powdered conductor with an insulator (WEISSHAAR & TALLMAN 1983), by impregnation of a porous conductor with an insulator (WANG ET AL. 1990), or by embedding carbon fibers in an insulating epoxy (DEUTSCHER & FLETCHER 1988). However, the commonly used types are array electrodes with geometries of microdisc and interdigitated microband.

3.2.3 Advantageous properties of microelectrodes

Compared to macroelectrodes, the attractive features of microelectrodes can be summarized as follows (PONS & FLEISCHMANN 1987, WIGHTMAN 1988, AOKI 1993):

- High mass transport rate (or mass flux) due to radial diffusion
- Rapid attainment of steady state, showing sigmoidal cyclic voltammogram and limiting current
- Capability of performing high-speed experiments, because of its significantly reduced charging current and small RC time constant
- Possibility of measurements in high resistance solutions, due to its extremely small current (i.e. the immunity to ohmic drop)
- High sensitivity thanks to the improved ratio of faradaic-to-charging current (I_F/I_C) and signal-to-noise ratio (S/N)
- Independent of convection
- No necessity of deoxygenation
- Increased current response with composite microelectrodes

3.2.3.1 Mass transport

3.2.3.1.1 Microspherical electrodes

For simplifying the studies, a microsperical electrode is considered. A potential step is applied to a microspherical electrode in a solution which contains only supporting electrolyte and electroactive species with a concentration C^{∞}. The potential is stepped from a value where no electrode reaction occurs to one where the electrolysis proceeds at diffusion controlled rate.

The concentration gradient at the electrode surface is obtained by solving Fick's second law in spherical coordinates, i.e.

$$\frac{\partial c}{\partial t} = D\frac{\partial^2 c}{\partial r^2} + \frac{2D}{r}\frac{\partial c}{\partial r} \qquad (1)$$

The initial condition is at $t = 0$ and $r > r_S$, $c = c^\infty$

The boundary conditions are for $t > 0$, at $r = \infty$, $c = c^\infty$ and at $r = r_S$, $c = 0$

where r is the distance from the centre of the sphere, r_S is the radius of the sphere, D is the diffusion coefficient of the electroactive species, and C is the concentration as a function of distance r, and time, t.

The equation can be solved using Laplace transform techniques and the current density I_d is given as follows:

$$I_d = \frac{nFDC^\infty}{r_s} + \frac{nFD^{1/2}C^\infty}{\pi^{1/2}t^{1/2}} \qquad (2)$$

where n is the number of electrons transferred in the redox reaction, F is Faraday's constant.

It can be seen from equation (2) that the current density contains a time-independent and time-dependent term. For a microelectrode, the response is the sum of a steady state and a transient component, both of which can be significant. Indeed, depending on the timescales, three types of diffusion regimes can be distinguished:

(i) Short times

At very short times, the second term of equation (2) will be so large that the first term can be ignored. The current density is given by the Cottrell equation, which is proportional to the reciprocal of the square root of time.

$$I_d = \frac{nFD^{1/2}C^\infty}{\pi^{1/2}t^{1/2}} \quad \text{(Cottrell equation)} \qquad (3)$$

The thickness of the diffusion layer is much smaller than the electrode radius. The mass transport process is dominated by linear diffusion as observed with macroelectrodes.

(ii) Long times

At a very long time, the second term of equation (2) can be negligible. The current density is merely inversely proportional to the radius of the electrode, indicating that the time-independent steady state current (i.e. limiting current) is attained and the radial or spherical diffusion prevails.

$$I_d = \frac{nFDC^{\infty}}{r_s} \qquad (4)$$

(iii) Intermediate times

In the case of the times between the two limits, the full equation (2) must be considered and the mass transport process is complicated.

3.2.3.1.2 Microdisc electrodes

For microdisc electrodes, it is generally recognized that the current response could be expressed using a modified equation of microspherical electrodes, despite the non-uniform flux of electroactive species to the disc surface and the tertiary current distribution. By the substitution of $r_s = \pi a/4$, where a is the radius of the disc (KAKIHANA ETAL 1981, AOKI & OSTERYOUNG 1981, OLDHAM 1981), the space averaged current density is given by

$$I_d = \frac{4nFDC^{\infty}}{\pi r} + \frac{nFD^{1/2}C^{\infty}}{\pi^{1/2}t^{1/2}} \qquad (5)$$

At short times the current density can be expressed as

$$I_d = \frac{nFD^{1/2}C^{\infty}}{\pi^{1/2}t^{1/2}} \qquad (6)$$

and at long times

$$I_d = \frac{4nFDC^{\infty}}{\pi a} \qquad (7)$$

Equation (7) is also defined as steady state current density for microdisc electrodes. The steady state current (i_{ss}) is the product of the current density I_d and the microdisc area, πa^2

$$i_{ss} = 4nFDC^{\infty}a \qquad (8)$$

For microdisc array electrodes with the number of individual electrodes of m, the steady state current is theoretically the sum of thc currents of individual electrodes provided the interelectrode distance is sufficient large and no shield effect is observed.

$$I_{ss} = 4mnFDC^{\infty}a \text{ (limiting current)} \qquad (9)$$

3.2.3.2 Reduced charging current

When an electrode is placed in an electrolytic solution, a double layer is formed at the interface. The electrical double layer resembles an ordinary capacitor. When the applied potential is changed, a current must flow to charge or discharge the double-layer capacitance. It is known as charging current (non-faradaic current), which is additive to the faradaic current and distorts the experimental data in the electron transfer process. For a potential step ΔE, the charging current i_c decreases exponentially with time at a rate dictated by the cell time constant (RC) (WIGHT MAN 1981, WANG 2000).

$$i_c = \frac{\Delta E}{R}\exp\left(\frac{-t}{RC}\right) \qquad (9)$$

where R is the cell resistance, C is the capacitance of the electrode, t is the time.

In an electrochemical cell, the resistance depends on the specific conductance of medium κ, and the electrode radius r_S (NEWMAN 1970).

$$R = \frac{1}{4\pi\kappa r_s}$$

It shows that cell resistance is conversely proportional to the electrode radius, i.e.,

$$R \propto \frac{1}{r_s}$$

Since the capacitance of the microelectrode is proportional to the electrode area,

$$C \propto r_S^2$$

Thus, $RC \propto \frac{1}{r_s} \bullet r_S^2 \propto r_S$

Therefore, RC decreases with decreasing electrode radius r_S and the charging current becomes less at all times.

3.2.3.3 Ohmic effect

When currents flow through a solution, they generate a potential that acts to weaken the applied potential on working electrodes by an amount of iR (where i is the total current, and R is the cell resistance). This can lead to severe distortions of experimental responses (ohmic effect). For conventional electrodes, the iR drop is usually minimized by electronic compensation. In contrast, the ohmic effect is significantly reduced with microelectrodes, because the faradaic currents become extremely small. As a result, electrochemical experiments could be performed in high resistive media using microelectrodes (BOND 1994, RYAN ET AL. 1994).

At short experimental timescales and in non-steady state conditions, where linear diffusion is dominant, the current is proportional to the electrode area (r_S^2),

$$i \propto \pi r_S^2 \propto r_S^2$$

While the cell resistance is given as

$$R = \frac{1}{4\pi\kappa r_s} \text{ or } R \propto \frac{1}{r_s}$$

Therefore,

$$iR \propto r_S^2 \bullet \frac{1}{r_s} \propto r_S$$

The iR drop decreases with decreasing the radius of microelectrode. On the other hand, in steady state conditions at long experimental timescales, the current response depends only on the radius, making the iR drop independent of the dimension of microelectrodes (BRUCKENSTEIN 1987).

4. Fabrication of ultramicroelectrode arrays

4.1 Introduction

Due to the rapid development of silicon technology and microelectronics, microelectrodes can be nowadays successfully fabricated using thick-film (screen-printing) (WRING & HART 1992, SEDDON ET AL. 1993, HART & WRING 1997) and thin-film (photolithography) techniques (FIACCABRINO & HOUDLKA-HEP 1998, SUZUKI 2000) in mass production, with well-defined and reproducible geometries, and at low costs (PALCHETTI ET AL. 2001). Thick film technology is called screen-printing technology, since the conventional screen-printing method is employed for the film deposition. The equipments needed in this technique are not complex and the microelectrodes produced in this method are inexpensive, being disposable after the measurement (i.e., single use). In contrast, thin film technology uses the modern but more expensive microlithographic technique, which is more expensive in the fabrication of microelectrode devices. Electrode structures as small as 0.5 µm can be fabricated using thin film technology (WANG ET AL. 1999), while only relative large structures, can be produced using screen-printing method, with characteristic feature size of ≥ 50 µm

(PRUDENZIATI 1994). The commonly used electrode materials are noble metals (e.g., Ir, Pt, Au), and carbon (FIACCABRINO & KOUDLKA-HEP 1997). However, for the mercury electrodes used in stripping analysis, it was found that Ir is the most suitable substrate, owing to its advantageous properties, such as low solubility in mercury ($<10^{-6}$ wt.%), no formation of intermetallic compounds with mercury and metal ions, good wettability by mercury, and high resistance to oxidation (KOUNAVES & BUFFLE 1986, GUMINSKI & GALUS 1986, KOUNAVES & BUFFLE 1987, SAMUEL & KOUNAVES 1991, KHAKANI ET AL. 1998). Incidentally, it should be mentioned that using thin-film and micromachining technologies, it is nowadays even possible to fabricate complete analytical microsystems on a chip (lab-on-a-chip), including fluid-handling silicon microstructures, on-chip three-electrode system, integrated potentiostat etc., which is very advantageous for in situ on-site analysis of heavy metals (REAY ET AL. 1996, HASWELL 1997, GUENAT ET AL. 1999, WANG ET AL. 1999, CAI ET AL. 2000, SCHULTZE & BRESSEL 2001).

In this study, three different UMEA chips (Ir/Pt, Pt, and Ir) with disc shaped electrodes of 0.75, 0.9, and 1.0 µm in radius, respectively, were tested. The chips were fabricated at ISIT using the lift-off and ion-milling of thin film technologies (UHLIG ET AL. 1995, UHLIG ET AL. 1997).

4.2. Fabrication of UMEAs

4.2.1 Design

On a silicon chip with a size of 6.4 mm x 5.0 mm four separate electrode arrays (working electrodes), two connected counter electrodes, and a small optional electrode are arranged (Fig. 9). Each array consists of 46 x 22 = 1012 disc shaped microelectrodes with an interelectrode distance of 25 µm. Three different chip varieties were designed with disc diameters of 1.5 µm, 1.8 µm, and 2.0 µm, which correspond to electrode areas of 1788, 2575, and 3179 μm^2 per array.

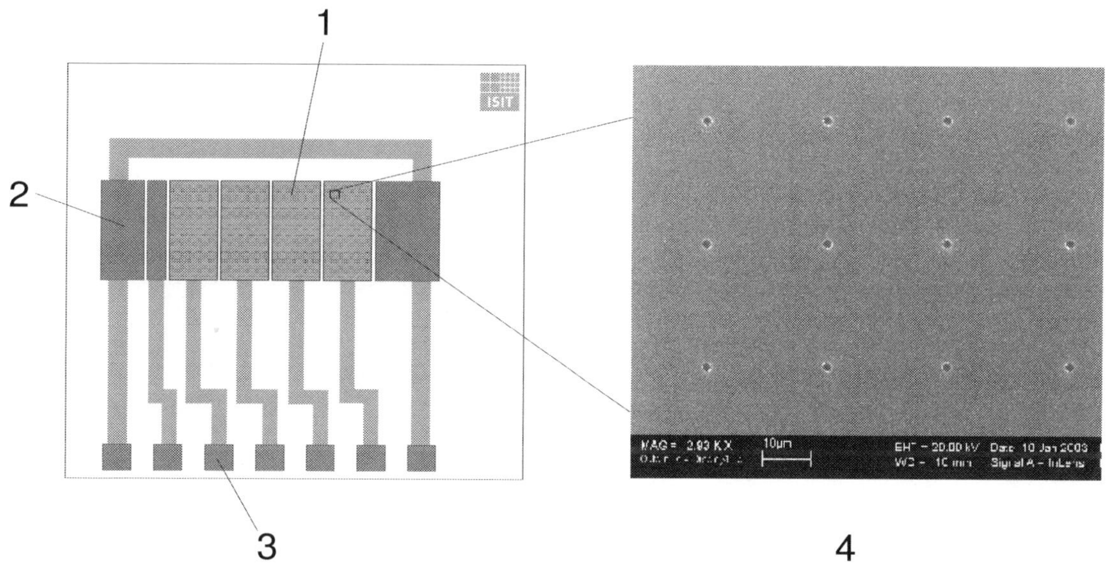

Fig. 9: Scheme of the ultramicroelectrode array (UMEAs) chip: 1 – microelectrode array, 2 – auxiliary electrode, 3 – contact pad, 4 – SEM picture

4.2.2 Processing

Figure 10 shows the process flow of the two alternative methods used in the fabrication of UMEAs at ISIT. As an insulating basis for the electrodes, thermal oxide (650 nm thick) was firstly grown on 6-inch silicon wafers. After that the electrodes are structured on the wafers by two different methods, i.e. lift-off and ion milling, depending on the type and thickness of the noble metal.

Figure 10a shows the lift off process. After a standard photolithography, titanium (20 nm thick) serving as adhesion layer is firstly evaporated, then the electrode layer (120 nm Pt or 70 nm Pt and 50 nm Ir) and connector layer are built. After a standard photolithography, the lift-off process was carried out in a special solvent (EKC) at a temperature of 85°C supported by ultrasonic bath, in order to remove the unexposed resist together with the unnecessary metal.

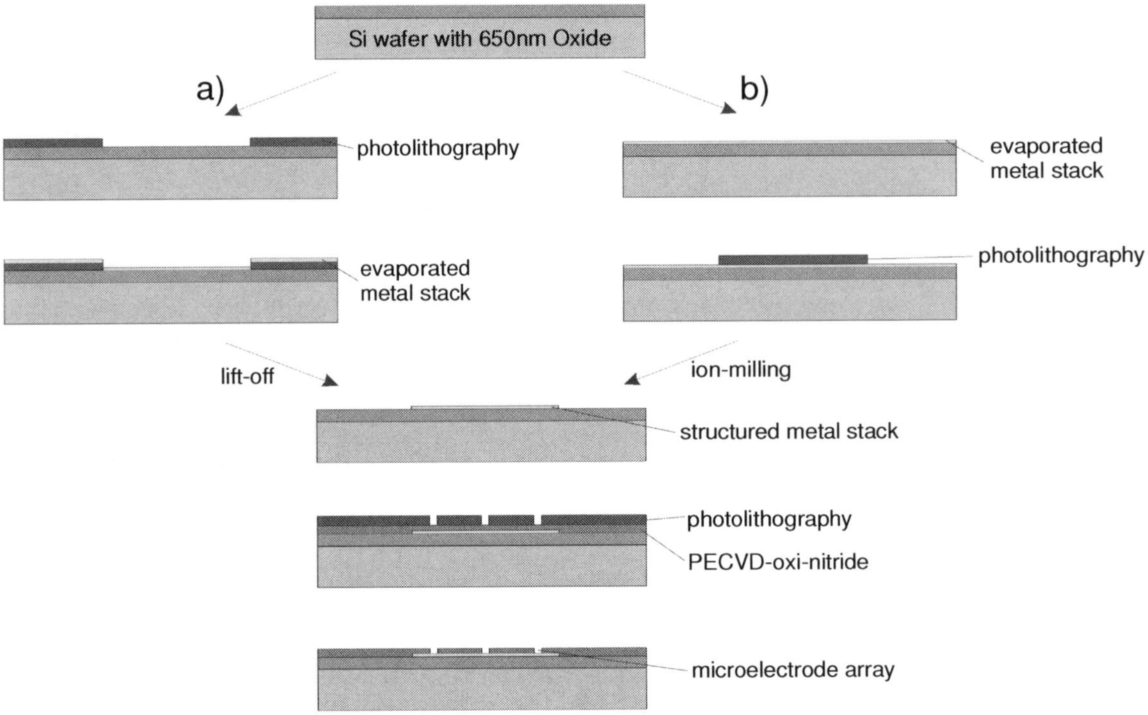

Fig. 10: Schematic process flow of the two alternative methods used in UMEA fabrication: (a) lift-off process, (b) ion-milling process

Since it is difficult to generate a thick iridium layer using the lift-off technique due to the hard and brittle property of iridium, in the second attempt (Fig. 10b) the ion-milling method was employed. The metal stack with 20 nm titanium and 200 nm iridium was firstly evaporated on the whole surface of the wafers; then, after a standard photolithography, the metal was structured by ion-milling, removing completely the metal between electrodes, pads and other structures.

After the metal stack was structured, the surface was covered by a passivation layer (500 nm silicon oxi-nitride), which is grown by plasma-enhanced chemical vapour deposition (PECVD-SiO_xN_x). The areas acting as working electrodes, auxiliary electrode, and connection pads were defined photolithographically and opened by dry etching. After some cleaning steps the chips are separated by sawing.

4.2.3 Assembly and packaging

The chips are mounted and bonded on printed circuit boards, which are electrically connected by inserting them in a commercially available 1/20-inch plug unit. The connection between chip and board is done by wire bonding using aluminum wires.

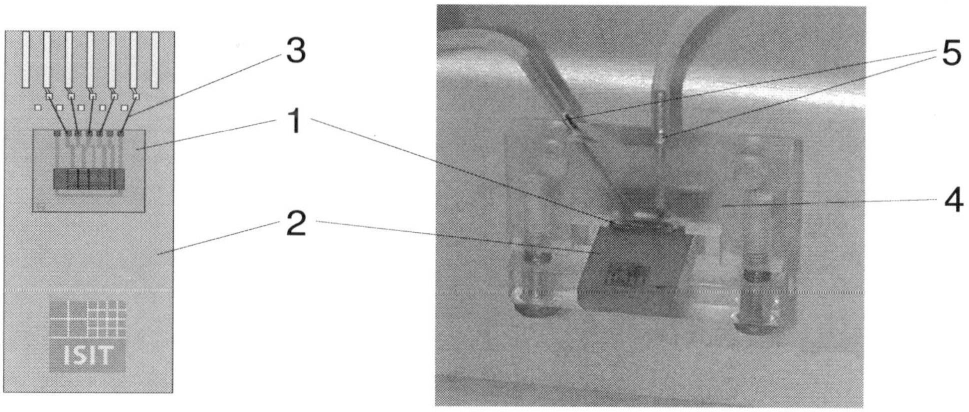

Fig. 11: Assembled and packaged UMEA chip: 1 – UMEA chip, 2 – printed circuit board, 3 – bond wires, 4 – flow through cell, 5 – stainless steel tubes for fluidic connection

The bond wires have been encapsulated in silicone. A flow through cell with a cavity of approximate 2-3 µl above the electrodes was sealed by means of an O-ring to the surface of the chip (Fig. 11). Two stainless steel tubes of the flow cell served as fluidic connectors to a peristaltic pump. The flow cell was constructed in a manner to allow replacing and multiple-use of UMEA chips.

5. Characterization of ultramicroelectrode arrays

5.1 Introduction

One of the most important goals in using microelectrode arrays is to increase the levels of measurement currents while maintaining at the same time the advantageous features of single microelectrodes. The significantly high collective currents obtained by using microelectrode arrays overcome the instrumental difficulties of measuring extremely small currents of single microelectrodes (CAUDILL ET AL. 1982, CAMMANN ET AL. 1991, MEYER ET AL. 1995). However, the condition is that the microelectrode arrays have an adequate design and are of good quality, including a sufficient interelectrode distance to avoid the shield effect due to the overlapping of diffusion layers (JU ET AL. 1992), without residues of the photoresists on electrodes, a good adhesion between the different layers, well isolated individual electrodes, etc. (FEENEY ET AL. 1997, FEENEY & KOUNAVES 2000). If microelectrode arrays are not properly developed, they will lose the advantageous properties, and will show the characteristic of macro- or quasi-microelectrodes. In order to check the quality of the UMEAs, several methods were used in this study.

5.2 Materials and methods

5.2.1 Apparatus and reagents

All electrochemical experiments were performed with an Autolab PGSTAT 10 potentiostat (Eco Chemie, Netherlands) in combination with the GPES software (Version 4.7) and a personal computer. A three-electrode configuration was realized by means of a micro-flow-through cell, UMEA-working electrodes, an external 16-702 micro-flow-through Ag/AgCl (3 M KCl) reference electrode (Microelectrodes, INC, USA), and an internal (on chip) iridium sheet counter electrode, or if necessary, an external stainless steel tube counter electrode. The micro-flow-through Ag/AgCl reference electrode is composed of an internal silver-silver chloride electrode with an internal filling solution of 3 M KCl saturated with AgCl. Compared to conventional reference electrodes, micro-flow-through reference electrodes are very small in size and can be used in a flow through mode during the experiment. A peristaltic pump MS Reglo (Ismatec, Switzerland), equipped with Tygon tubing was used for propulsion of the solution to the micro-flow-through cell. Images of the UMEAs were taken with CAMECA S 50 (France) scanning electron microscope (SEM) and Philips XL 30 FEG environmental scanning electron microscope (ESEM). Surface morphology of microelectrode was recorded using atomic force microscope (AFM)

(Dimension 3100, Digital Instruments, USA) with NanoScope IV Control software. Pictures of mercury film were taken by means of optical microscope (Zeiss, Germany) with digital image analyser (DIANA) software and Sanyo colour CCD digital camera. For cyclic voltammetry (CV) and chronoamperometry (CA), a solution composed of 6mM potassium hexacyanoferrate (III) ($K_3Fe(CN)_6$) (Merk, analytical grade) in 1 M potassium nitrate (Merk, analytical grade) was used. Milli-Q water was used for the preparation of all solutions.

5.2.2 Methods used for characterization of the UMEAs

The UMEA chips were characterized by surface analysis and electrochemical methods. Prior to each electrochemical experiment, the quality of the chip was investigated with respect to morphology, surface structure, and chemical composition by means of optical microscope; followed by scanning electron microscope (SEM), environmental scanning electron microscope (ESEM), and atomic force microscope (AFM). The advantage of ESEM lies in the possibility of examining electrode surface without coating with carbon or gold, which would preclude the further use of the chips in the electrochemical experiments. Before use, the chip was firstly conditioned by cycling the potential several times in the range of -0.6 and 1.3 V at a sweep rate of 10 mV/s in a 0.8 N H_2SO_4 solution using cyclic voltammetry. Afterwards, the chip was characterized by cyclic voltammetry in the potential range of +0.9 and -0.3 V using different sweep rates, and finally, by chronoamperometry at -0.6 V in a 6 mM $K_3Fe(CN)_6$ /1 M KNO_3 solution. The experimental set up is schematically illustrated in figure 12. Experimental solutions were pumped through the micro-flow-through cell mounted on the electrode surface of the UMEA chip, through the external counter electrode, the micro-flow-through reference electrode, and then to waste.

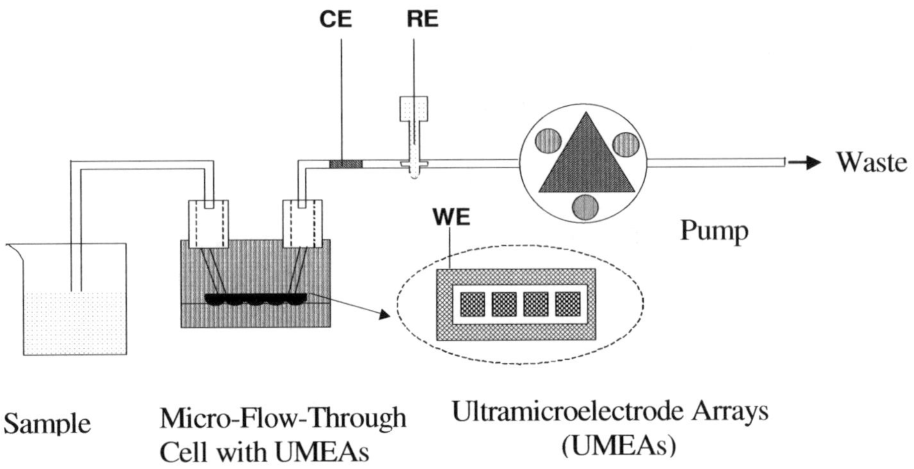

Fig. 12: Experimental set up

5.3 Results and discussion

5.3.1 Structure of the Ir-UMEAs

The microstructure of UMEAs obtained as revealed by SEM and AFM is shown in figures 13a ~ f. A very regular arrangement of arrays and electrodes are shown in figures 13a and 13b. The diameter (d) was measured to be 1.825 µm and 1.762 µm, respectively; and the interelectrode distance of UMEs (w) was measured to be 25.45 µm and 24.56 µm, respectively, which are in very good agreement with data given by the fabrication (d = 1.80 µm and w = 25.00 µm) (Fig. 13c and 13d). A three dimensional image of the UME is shown in figures 13e and 13f. Instead of an inlaid electrode, the picture in figure 13e reveals a recessed electrode with a depth of L = 0.2 µm. The morphology of UMEs was analyzed by AFM. The roughness index of the UME was found to be 1.0 ± 0.2 nm, which indicates a relative smooth surface (Fig. 13f). This manifests that the iridium layer used as electrode material was well evaporated during the fabrication of the UMEAs, which is considered to be an important parameter for the voltammetric measurements (STUKIK 1992).

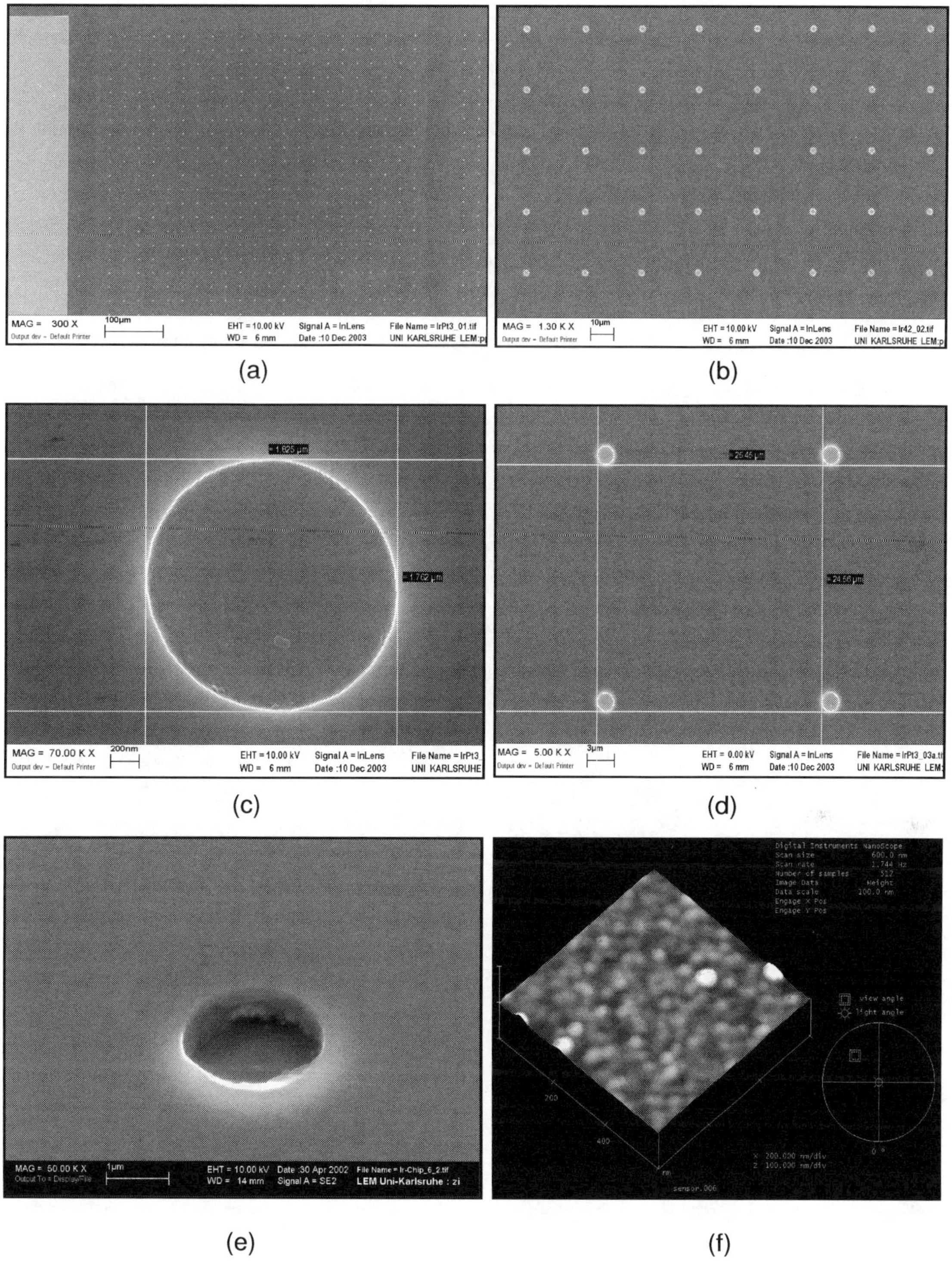

Fig. 13: SEM and AFM images of UMEA, (a) magnification of 300 X, (b) magnification of 1300 X, (c) diameter of UME, (d) interelectrode distance, (e) recess depth, (f) surface roughness of UME

5.3.2 Surface analysis

Figures 14a and 14b show two examples of ESEM pictures of iridium UMEs. Very regular disc shaped electrodes with well-defined areas and sharp edges are observed, which are typical features for the microdisc geometry. The surfaces of the UMEs are totally free of photoresist deposition or passivation layer. However, at large magnification (40,000 X), fine cracks could be found on the surfaces of all UMEs, which may be due to the difficulty in generating a thick Ir film of iridium during the fabrication process (WANG ET AL. 2000). Because iridium is a very hard and brittle noble metal (KHAKANI ET AL. 1997, WECHTER & OSTERYOUNG 1990), it is difficult to be used to generate an electrode layer thicker than 200 nm, even by means of ion-milling technique (FIACCABRINO & KOUDLKA-HEP 1997). Nevertheless, it was found that such slight cracks have no influence on the electrochemical experiments.

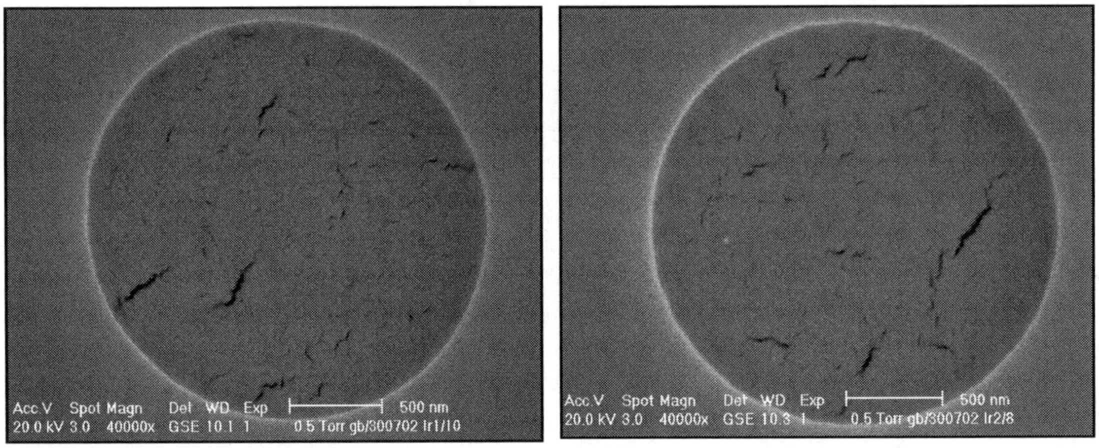

(a) (b)

Fig. 14: ESEM pictures of Ir UMEs

Figures 15a and 15b indicate ESEM pictures of Ir/Pt-UMEs. In contrast to the Ir UMEs, residues of photoresist or passivation layer could be frequently found on the surface of the Ir/Pt-UMEs, indicating that cleaning was incomplete during the chip fabrication. As shown in SEM pictures in figures 15c and 15d, the surface of the Ir/Pt-UMEs may be almost completely covered by isolator materials, probably blocking the UMEs and preventing them to work properly, although this seldom occurs. However, a mixture of concentrated H_2SO_4 and 10% H_2O_2 proves to effectively remove the residues and to clean the Ir/Pt chip, as it was confirmed by cyclic voltammetry during the characterization of the chip.

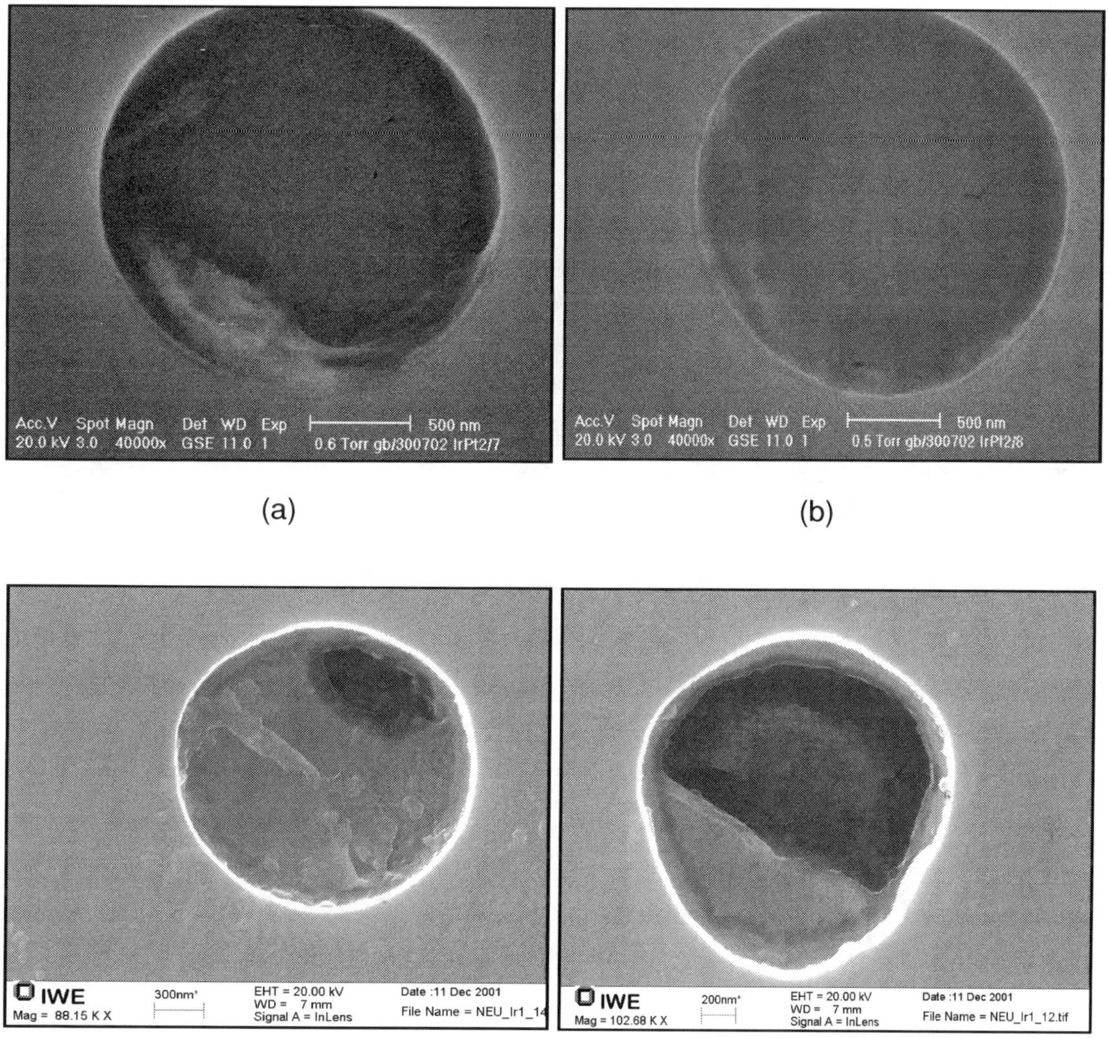

(a) (b)

(c) (d)

Fig. 15: ESEM and SEM pictures of Ir/Pt UMEs

As shown in the ESEM pictures in figure 16, the electrode surfaces of Pt-UMEA chips have grainy texture. Although the ESEM pictures do not reveal fabrication failure, electrochemical experiments such as cyclic voltammetry could not be carried out, which suggest that Pt-chips are of a poor quality. However, the exact reason is not clear and further investigation is necessary in this respect. Indeed, the disadvantages of Pt-based Hg-plated microelectrode sensor for the measurement of trace metal concentrations, such as the relative large solubility of Pt in Hg (GUMINSKI & GALUS 1986, KOUNVES & BUFFLE 1986, KOUNVES & BUFFLE 1987), the formation of intermetallic compounds with Cd and Zn (e.g., Pt-Cd and Pt-Zn) (KOUNAVES & BUFFLE 1986, BALDO ET AL. 1995,) is well known. Therefore, the fabrication of Pt-based UMEA sensor is less important.

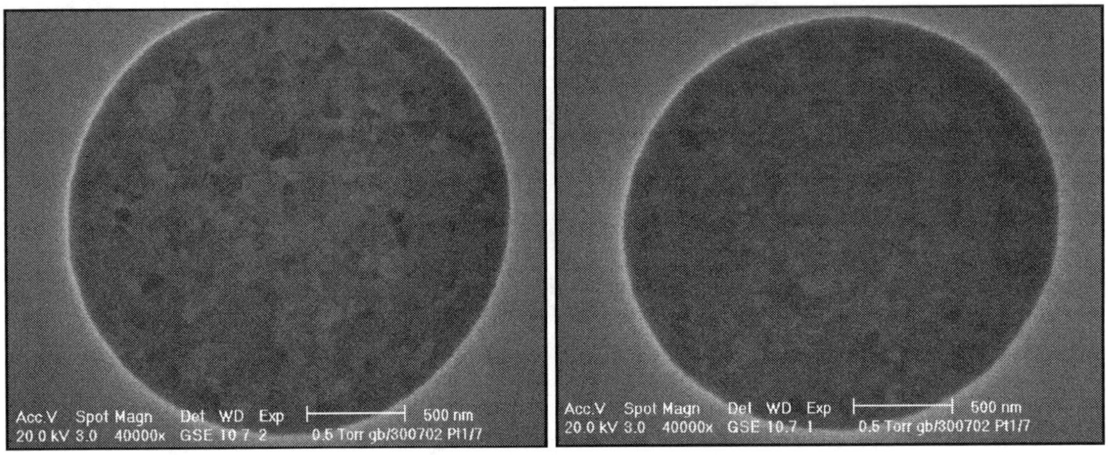

(a) (b)

Fig. 16: ESEM pictures of Pt-UMEs

5.3.3 Electrochemical behavior

Cyclic voltammetry (CV) and chronoamperometry (CA) are the commonly used methods for the characterization of microelectrode arrays.

The cyclic voltammetric behavior of individual microelectrodes is well known, i.e., at slow scan rates, the shape of the cyclic voltammogram is sigmoidal (WIGHTMAN 1981, HEINZE 1981, WIGHTMAN 1988), indicating that the steady state (BOND ET AL. 1989, ZOSKI 1990, ZOSKI & BOND 1990) is attained due to the radial diffusion and high mass transport rates of microelectrodes; whereas at high scan rates, a peak shaped cyclic voltammogram results (HEINZ 1993, CHING ET AL. 1994, BUCHBERGER 1998). However, the electrochemical behavior of microelectrode arrays is more complicated. Depending on the way the microelectrodes are arranged in arrays (geometry) or on the scan rates of experiments (timescales), the same array may exhibit the cyclic voltammetric behavior of micro-, quasi micro-, or macroelectrodes.

For a given experimental timescale, the arrangement of microelectrodes (geometry) plays a decisive role in the electrochemical behavior of microelectrode array. For a loosely packed array with a sufficient large spacing between electrodes, it generates a multiple current response of individual electrodes, retaining all of the advantageous properties of microelectrodes, such as high mass transport of radial diffusion and steady state behavior (MORF & ROOIJ 1997, MORF 1997). A sigmoidal shaped cyclic votammogram will arise. In contrast, for high densely packed microelectrode array, the diffusion layers of individual electrodes will totally overlap, and the array shows a cyclic behavior of macroelectrodes, for instance, a peak shaped cyclic voltammogram. As for microelectrode array with a moderate packed density, the diffusion layers of individual electrodes will partially overlap, leading to a mixed mass transport regime (Lee et al. 2001). As a result, the array will possess a peculiar electrochemical feature (ZHANG 2000). Actually, the optimisation of microelectrode array is to find a geometry, which tends towards unity so that the array behaves as a set of independent microelectrodes, but for which the number of microelectrodes per unit area is as large as possible in order to increase the current signal (SEDDON ET AL. 1994).

The timescale or scan rate is a very important factor for the electrochemical behavior of microelectrode array with a definite geometry. Particularly, interactions among the diffusional fields of adjacent microelectrodes have to be considered. Generally, three different cases can be distinguished (Fig. 17) (RELLER ET AL. 1984, PENNER & MARTIN 1987, CHENG & MARTIN 1988). At very short times or very high scan rates, linear diffusion dominates within individual electrodes and a broad peak shaped cyclic voltammogram curve results, indicative of the Cottrel state (BARD & FAULKNER 1980). Similar effect is also observed at very long times or very slow scan rates, when linear diffusion is restored due to the overlap of the diffusion layers of individual electrodes. The main difference is that at short times the linear diffusion involves only the electrically active area of microelectrode array (PENNER & MARTIN 1987) and consequently the current is proportional to the area of electrodes; whereas at long times, the linear diffusion concerns with the total geometric area of the microelectrode array and the current is proportional to the total area of array (CHENG & MARTIN, AMATORE 1995). In both cases, the current is inversely proportional to the square root of time ($t^{-1/2}$).

$$i = \frac{nFDCA_{el}}{(\pi Dt)^{1/2}} \quad \text{(at very short times)}$$

$$i = \frac{nFDC(A_{el} + A_{insul})}{(\pi Dt)^{1/2}} \quad \text{(at very long times)}$$

Where A_{el}, the area of the electrodes (active area); A_{insul}, the nonactive surface area; n, the number of electrons transferred in a redox reaction; F, the Faraday's constant; D and C, the diffusion coefficient and the concentration of the electroactive species, respectively; and t, the experimental time.

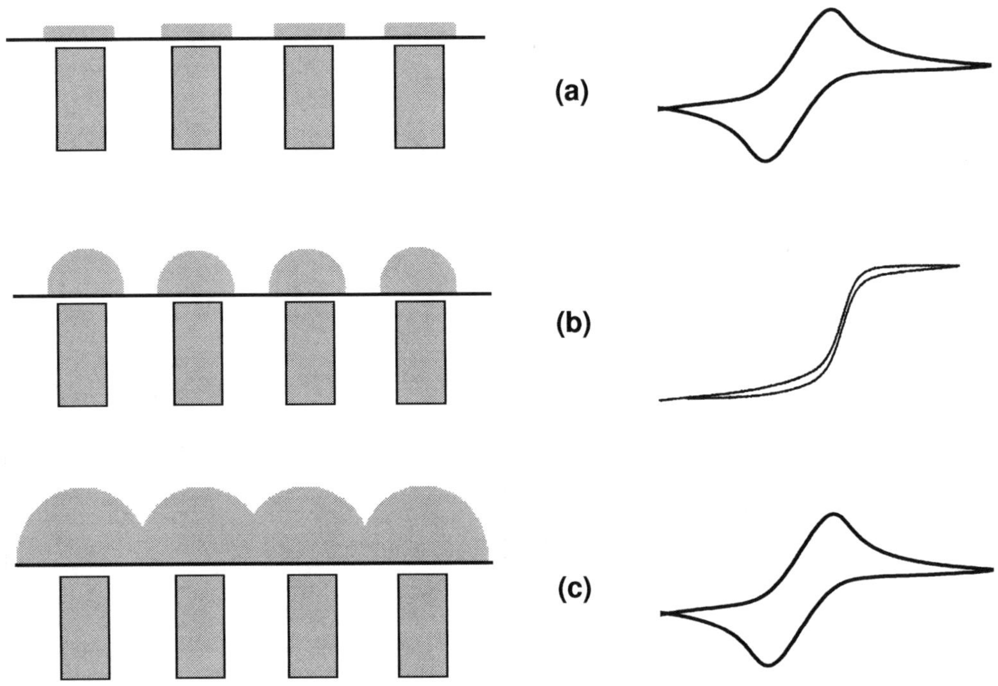

Fig. 17: Development of diffusion concentration profiles in microelectrode array and the corresponding cyclic voltammograms: (a) short times or high scan rates, (b) intermediate times or scan rates, (c) long times or slow scan rates

Interestingly, at very long timescales, the signal-to-noise ratio of microelectrode array is much improved, because the signal is proportional to the whole surface of the array, while the noise is proportional only to the active area of the array (i.e., the area of electrodes) (STULIK ET AL. 2000).

As for intermediate times or scan rates, the linear diffusion dominant at short times transits to radial diffusion (or non linear diffusion), and consequently, a steady state will be attained, with a characteristic sigmoidal cyclic voltammogram The current is simply the sum of steady state currents of individual electrodes:

$i = i_{ss} \times m = 4\ m\ n\ F\ D\ C\ r$ (at intermediate times)

where i_{ss} is the steady state current of an individual microelectrode, r is the radius of electrode, m is the number of microelectrodes.

In general, the achievement of steady state is very important for microelectrode array. It depends mainly on three factors: (a) the size of the electrodes; (b) the duration of the experiment; and (c) the diffusion coefficient of the electroactive species. In principle, the establishment of steady state requires that $Dt/r^2 \gg 1$ and that the time is within the order of a few r^2/D (CHING ET AL. 1994, AMATORE 1995). For a given microelectrode array, the steady state looks like to be sandwiched between two Cotrellian relaxations (FLETCHER & HORNE 1999), as shown schematically in figure 17.

5.3.3.1 Cyclic voltammetric behavior

In this study, the cyclic voltammetric behavior of various UMEA chips (Ir-, Ir/Pt, and Pt-UMEA chips) was characterized using cyclic voltammetry (CV).

5.3.3.1.1 Ir UMEAs

The cyclic voltammetric behavior of the newly developed iridium UMEAs is shown in figure 18. Scan rates of cyclic voltammetry vary from 5 mV/s to 30 V/s.

At slow scan rates or long timescales (e.g., 5 mV/s and 10 mV/s, 20 mV/s, 50 mV) the UMEAs display a slight peak shaped cyclic voltammogram, which is indicative of the transition of the mass transport from radial diffusion to linear diffusion (SANDISON

ET AL. 2002). The slight decrease in measurement currents observed at a scan rate of 5 mV/s provides an evidence of the onset of linear diffusion. However, under such conditions the steady state component is still dominant (Fig. 18a ~ d).

At intermediate scan rates or moderate timescales (e.g., from 100 mV/s to 1000 mV/s), sigmoidal cyclic voltammograms result, suggesting that the individual diffusion fields of the electrodes remain isolated within the given timescale, and that steady state is attained radial diffusion due to radial diffusion (Fig. 18e ~ h). Nearly constant measurement currents (i.e., limiting currents) are observed, proving that at steady state the current response is independent of time or scan rate.

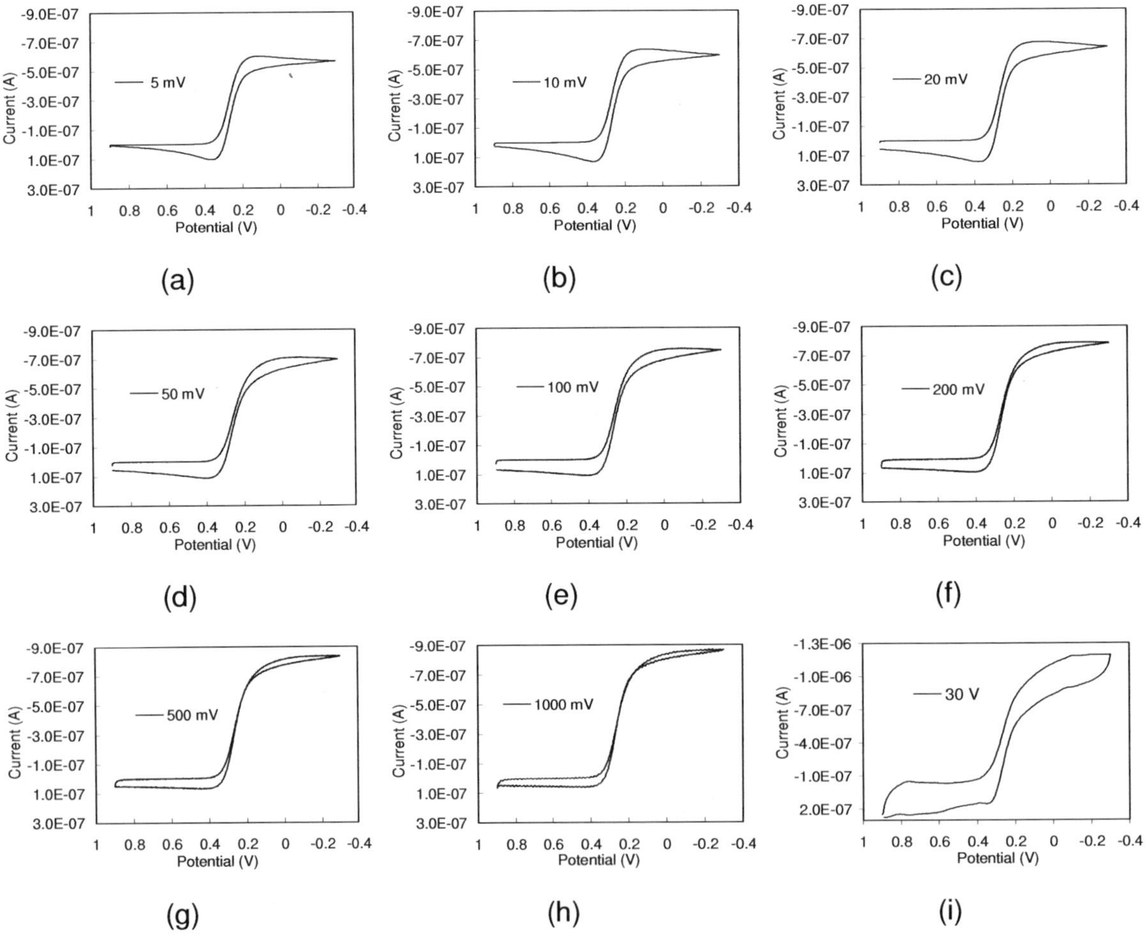

Fig. 18: Cyclic voltammograms obtained with a single Ir-UMEA in a solution of 6 mM $K_3Fe(CN)_6$ and 1 M $NaNO_3$ in a potential range from 0.9 V to –0.3 V, using different scan rates (a) 5 mV/s, (b) 10 mV/s, (c) 20 mV/s, (d) 50 mV/s, (e) 100 mV/s, (f) 200 mV/s (g) 500 mV/s, (h) 1000 mV/s, (i) 30 V/s

At high scan rates or short times (e.g., 30 V/s), the UMEA displays a deformed and slightly peak shaped cyclic voltammogram, and enhanced measurement current, showing the transition from radial diffusion to linear diffusion (Fig. 18i).

A typical peak shaped cyclic voltammogram as shown in figure 17 cannot be obtained at high scan rates or short times. Such a cyclic voltammetric behavior is most likely the consequence of planar diffusion confined to individual microelectrodes due to the very small radius of the microelectrodes (0.9 µm). The radial diffusion of microelectrodes is so strong that it is able to keep up with the consumption of ferricyanide at the electrode surface. It is documented that for very small electrodes a high scan rate is needed to see any behavior other than the steady state. For example, for an electrode of 0.5 µm in radius, steady state behavior would be hold up to 10 V/s (diffusion coefficient of electroactive species: $D = 10^{-5}$ cm^2/s, $T = 298$ K) (BARD & FAULKNER 2001). Incidentally, the slightly elevated current at a scan rate 30 mV/s may be probably explained by the Randles-Sevcik theory. In a mass transport regime of typical linear diffusion, for instance, in the case of macroelectrodes or conventional electrodes, the current response for a reversible chemical process increases proportionally to the square root of the scan rate (MONK 2001).

5.3.3.1.2 Ir/Pt UMEAs

The cyclic electrochemical behavior of Ir/Pt-UMEA chips is shown in figure 19. For chip labelled with Ir/Pt 1 the measurement current increases substantially (approx 23 times) after treatment with H_2SO_4, indicating an effective elimination of isolator residues (Fig. 19a, 19b). The nearly identical shape of voltammograms obtained before and after treatment with H_2SO_4 suggests that only a small number of electrodes were active in the experiment and were well cleaned in the fabrication process. For chip Ir/Pt 2, without treatment with H_2SO_4 a deformed voltammogram with a very small current was recorded, indicating a very poor cleaning during fabrication (Fig. 19c). After treatment with H_2SO_4, a cyclic voltammogram was obtained with a significant improvement in both measurement current and the shape of curve (Fig. 19d). Moreover, a satisfactory mercury film could be generated, allowing the measurement of trace concentrations of heavy metals by SWASV. Consequently, residues of isolator materials (e.g., photoresist, passivation layer) on the electrodes of the Ir/Pt chip proved to be less a problem in the electrochemical experiments.

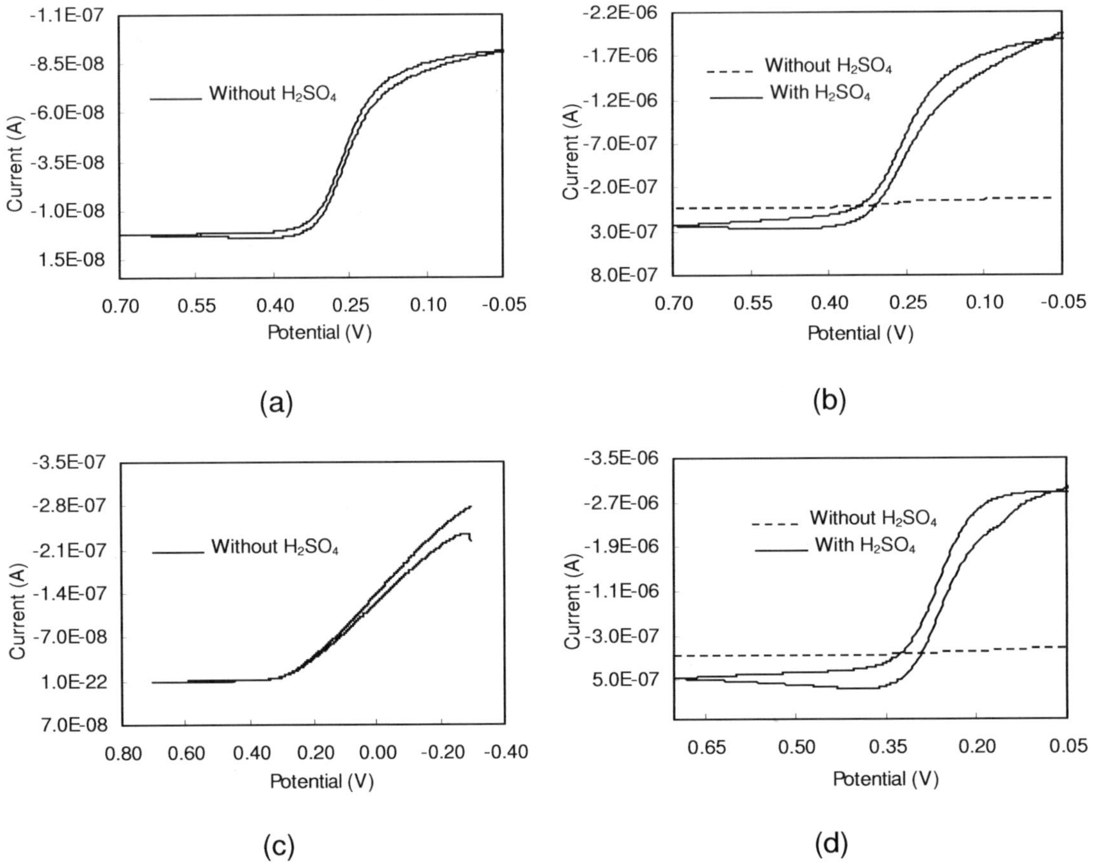

Fig. 19: Cyclic voltammograms obtained with an Ir/Pt-UMEA in a solution of 5 mM $K_3Fe(CN)_6$ and 1 M $NaNO_3$ at scan rate 20 mv/s. (a) Ir/Pt 1, before treatment with H_2SO_4, (b) Ir/Pt 1, after treatment with H_2SO_4, (c) Ir/Pt 2, before treatment with H_2SO_4, (d) Ir/Pt 2, after treatment with H_2SO_4

5.3.3.1.3 Pt UMEAs

Compared to Ir- and Ir/Pt UMEA chips, Pt UMEA chips show a much poorer cyclic voltammetric behavior (Fig. 20). There is no improvement in the voltammograms after the Pt chips were treated with concentrated H_2SO_4, both in curve shapes and in measurement currents. The use of other chemicals, such as concentrated HNO_3, acetone, etc. showed also to be ineffective, possibly suggesting fabrication failures.

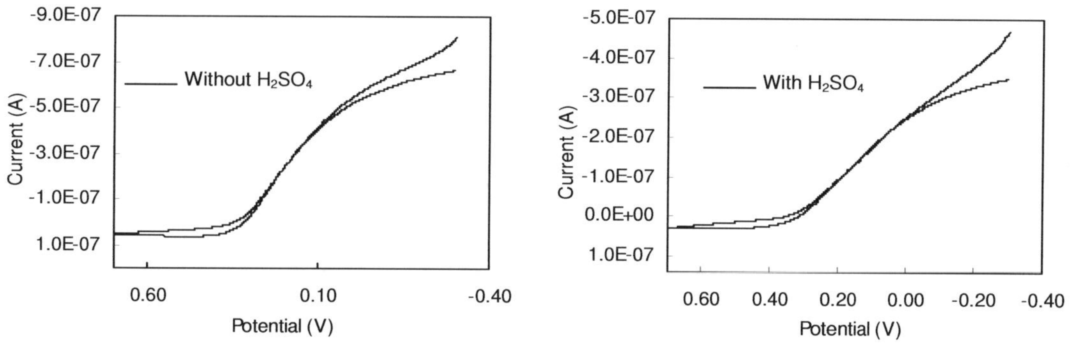

(a) (b)

Fig. 20: Cyclic voltammograms obtained with a Pt-UMEA in a solution of 5 mM $K_3Fe(CN)_6$ and 1 M $NaNO_3$ at a scan rate of 20 mv/s: (a) before treatment with H_2SO_4; (b) after treatment with H_2SO_4

5.3.3.1.4 Macroelectrodes

For comparison, the cyclic voltammetric behavior of a macroelectrode is shown in figure 21a. For recording of the cyclic voltammogram, the sheet counter electrode of the Ir chip (with a total area of 1.5 mm^2) was used as working electrode

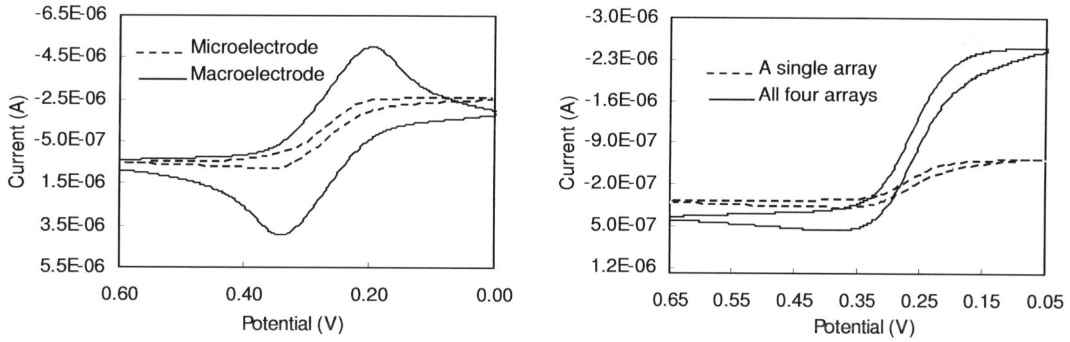

(a) (b)

Fig. 21: Comparison of cyclic voltammograms obtained with (a) a macroelectrode and an Ir-UMEA; (b) a single array and four arrays interconnected in a solution of 6 mM $K_3Fe(CN)_6$ and 1 M $NaNO_3$ at a scan rate of 20 mv/s

(macroelectrode), whereas as counter electrode the external stainless steel tube was used. Unlike UMEAs, the macroelectrode exhibits a classical peak shaped and broad cyclic voltammogram, with an enhanced measurement current. The different cyclic voltammetric characteristics are used to distinguish between macroelectrodes and microelectrodes.

5.3.3.1.5 Comparison of a single array with all four arrays interconnected

The cyclic voltammetric behavior of a single UMEA and all four UMEAs is compared in figure 21b. The measurement current using the four UMEAs jointly (3.06×10^{-6} A) is roughly 4 times the measurement current obtained with a single UMEA (7.80×10^{-7} A), indicating a multiple current response. The much higher measurement currents obtained with multiple arrays is very advantageous in electrochemical experiments, because they allow the use of conventional instrumentation, without the need for special low-noise amplifiers.

5.3.3.2 Amperometric response

Chronoamperometry (CA) records the change of the measurement current with time. As discussed before, at sufficient short time as well as long time the microelectrode array shows Cottrell behavior and the current response is inversely proportional to the square root of time ($t^{-1/2}$). Between the two limiting cases, microelectrode array reveals a behavior of steady state or quasi steady state (i.e., a mixed mass transport regime). Figure 22 shows the current response of a single UMEA in the time frame of 0.001 and 1000 s in a 6 mM $K_3Fe(CN)_6$ /1 M KNO_3 solution (the diffusion constant of $K_3Fe(CN)_6$ is 7.84×10^{-10} m^2 s^{-1}) and at a potential of –0.6 V. The diagrams indicate roughly two diffusion states, i.e., Cottrell state, and steady- and quasi-steady state. The Cottrell state occurs at a very short time of about 0.03 s (Fig. 22a and appendix 4), which is proved by the current decay with the reciprocal of the square root of time ($t^{-1/2}$) (Fig. 22b) (ANDREWS & HARRIS 1998). After approximate 10 s the current falls to a constant value, suggesting that the steady state was attained (Fig. 22c), which is time independent and is merely proportional to the radius of the microelectrode and

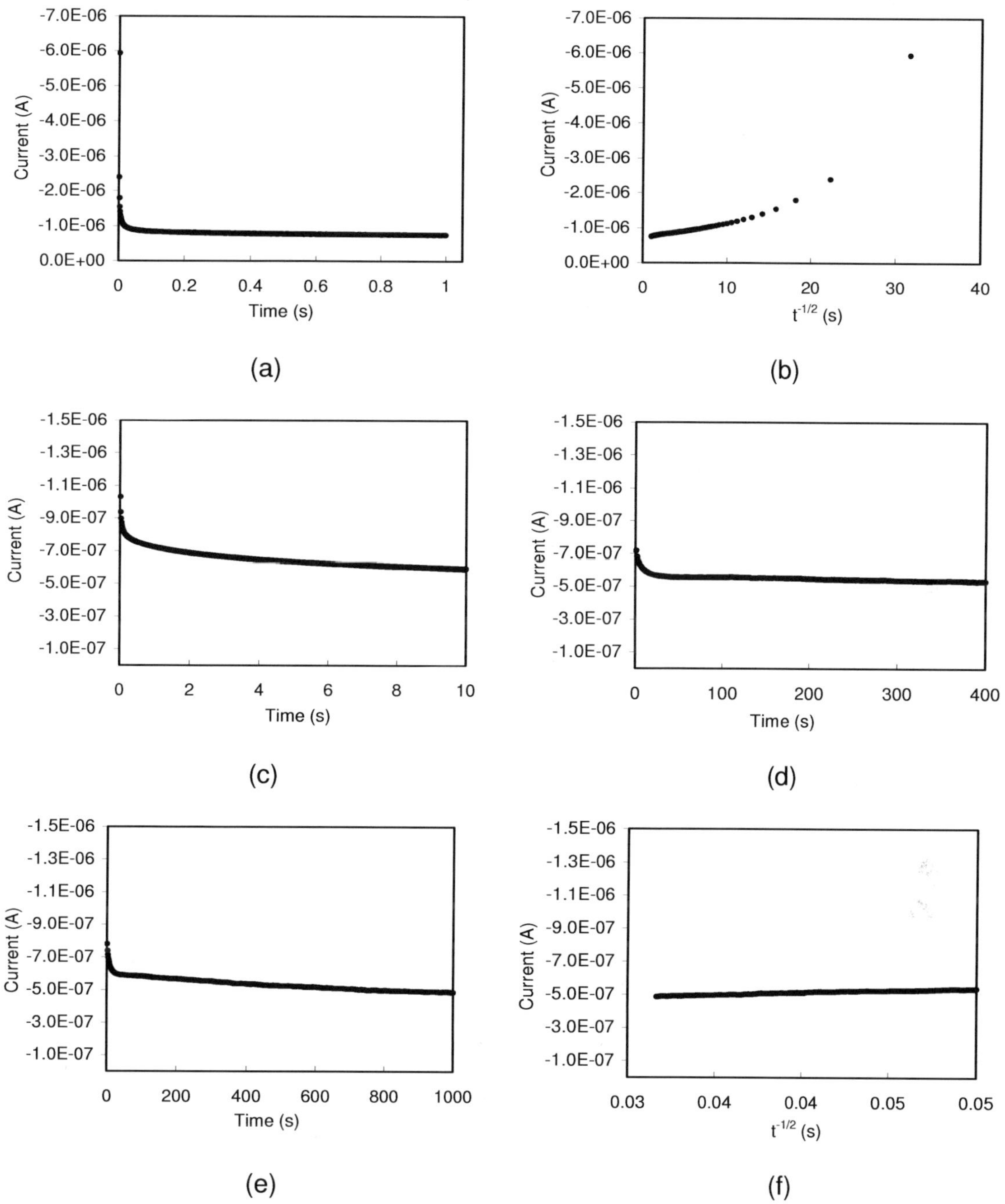

Fig. 22: Plots of the chronoamperometric response of Ir-UMEA, (a) current vs. t in 1 s; (b) current vs. $t^{-1/2}$ in 1 s; (c) current vs. t in 10 s; (d) current vs. t in 400 s; (e) current vs. t in 1000 s; (f) current vs. $t^{-1/2}$ from 400 to 1000 s. The experiments were performed in a solution of 6 mM $K_3Fe(CN)_6$ and 1 M KNO_3 at a potential of -600 mV

the concentration of metal ions. The pure steady state is maintained for about 400 s (Fig. 22d), and then a slightly decrease in current response is observed, revealing the onset of a linear diffusion component (due to the overlap of the diffusion layers of adjacent UMEs), and the transition from steady state to quasi steady state. However, for a time of 1000 s, the second Cottrell state is not attained, which is indicated by the nearly constant current response (Fig. 22e and appendix 5) and the absence of Cottrellian current behavior similar to the case at the very beginning of the test (Fig. 22f). If the time gets sufficiently longer, the typical Cottrell behavior will occur again. It should be noted that the chronoamperometric performance of UMEAs is in good agreement with their cyclic voltammetric behavior.

5.3.4 Simulation of current response

One of the most commonly used digital simulations for the chronoamperometric response of microelectrode array is from Shoup and Szabo (1984). According to the theory, the current response of a hexagonal array of microdisc electrodes with an inlaid geometry depends on the experimental time and the array geometry as follows:

$$i_{inlaid} = 4mnFDCr\, f(\tau,\theta) = i_{limiting} f(\tau,\theta) \qquad (1)$$

where $f(\tau,\theta)$ is the function of the experimental time and the geometry of the microelectrode array.

$$f(\tau,\theta) = \frac{\sqrt{\pi}}{2\sqrt{\tau}} + \frac{\pi}{4} + \left(1-\frac{\pi}{4}\right)\exp\left[-\frac{0.7823\theta^2(3-2\theta)}{\sqrt{\tau}}\right]$$
$$-\exp\left\{\frac{-\left[\frac{\sqrt{\pi}}{2}\frac{\theta}{1-\theta} + 0.7823\left(1-\frac{\pi}{4}\right)\theta^2(3-2\theta)\right]}{\sqrt{\tau}}\right\}\left(1+\frac{3}{2}\frac{\sqrt{\pi}}{\tau}\frac{\theta^3}{1-\theta}\right)^{-1} \qquad (2)$$

where $\tau = \frac{4Dt}{r^2}$, $\theta = 1-\left(\frac{2r}{d}\right)^2$, d is the interelectrode distance

In addition, according to Bond (1988) for recessed microelectrodes with a recess depth L, the current response of inlaid microelectrodes should be modified as:

$$i_{recessed} = i_{inlaid}\left[\frac{\pi r}{4L + \pi r}\right] \quad (3)$$

In this study, chronoamperometry was carried out in a 6 mM $K_3Fe(CN)_6$ solution at a potential of –0.6 V and a experimental time of 400 s. The parameters used are as follows:

m –the number of electrodes in a single array: 1012

n –the number of electrons exchanged: 1

F – the faradiac constant: 9.6485×10^4 c mol^{-1} i.e., 9.6485×10^4 A. s

D –the diffusion constant of $K_3Fe(CN)_6$: 7.84×10^{-10} $m^2 s^{-1}$

C – the concentration of $K_3Fe(CN)_6$, 6 mM, i.e., 6 mol/m^3

r – the radius of microelectrodes, 9×10^{-7} m

d – the interelectrode distance, 2.5×10^{-5} m

L – the recess depth of microelectrodes, 2×10^{-7} m

θ – the active fraction of surface area, 0.9948

t – time, 400 s

The comparison of the chronoamperometric results with the digital simulation (Fig. 23 and appendix 6) indicates a large difference in the current response. Compared to the theoretical values, the experimentally measured currents are much larger and more stable. It is difficult to give the exact reasons for the difference. Nevertheless, the following influence factors may be potentially considered: (i) the theoretical simulation model is not suitable for the UMEAs developed in this study. Theoretically, a good agreement can only be obtained if the fraction of the active surface area is in the range of $0.368 < \theta < 0.841$, meaning that the model is valid only for very densely packed microelectrode arrays (SCHARIFKER 1988, BERIET ET AL. 2000). In contrast, the UMEAs used in this study have a θ value of 0.9948, indicating a much more loosely packed geometry. Consequently, the discrepancy between the model and measured current response is possibly due to the inadequacy of Shoup and Szabo's empirical expression for high θ values, (ii) The high measurement currents may be due to the use of a too negative potential (-0.6 V) in chronoamperometry, which would cause the reduction of other compounds, such as oxygen, adding an auxiliary current to the measurements. However, by repeating the experiments at a more positive potential of -0.2 V in deoxygenated solution using conventional electrochemical cell and large reference electrode, almost no change in current response was observed. In fact, high and steady measurement currents for long experimental timescales which are very desirable properties of microelectrode arrays can be only attained for chips with adequate geometry and good fabrication quality. For simulation of chronoamperometric response of the UMEAs used in this study, alternative theoretical models should be considered.

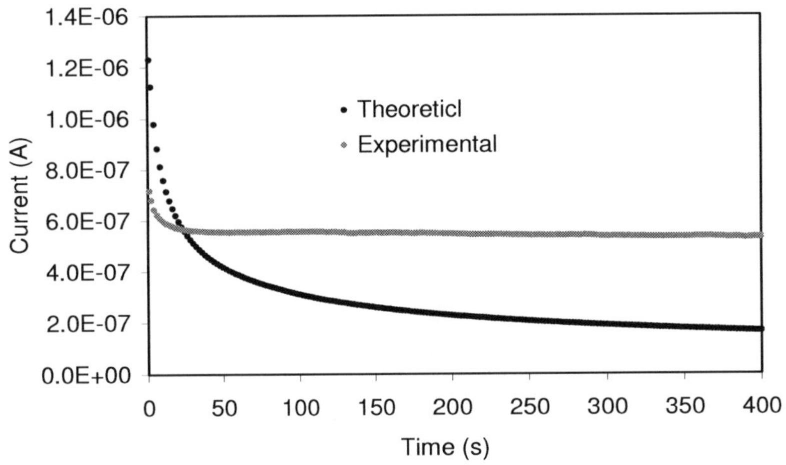

Fig. 23: Comparison of the experimental current responses with the theoretical values obtained with the digital simulation of Shoup and Szabo (1984). The experiments were carried out in a solution of 6 mM $K_3Fe(CN)_6$ and 1 M KNO_3. Potential was stepped from +500 mV to –600 mV for a timescales of 400 s

5.4 Conclusion

Compared to Ir/Pt- and Pt-UMEA chips, the Ir-UMEA chips used in this study show satisfactory performances in respect of surface quality, cyclic voltammetry, and choronoamperometry. The SEM and ESEM analysis indicate that the Ir chips are of good quality (regular discs, well-defined area, no residue of passivation layer or photoresist etc.). Instead of broad and planar CV peaks with macroelectrodes, sigmoidal cyclic voltammograms are observed with the Ir-UMEAs at scan rates varying from 5 mV/s to 30 V/s, indicative of the properties of microelectrodes. On the other hand, the slightly peak shaped voltammograms at slow scan rates (e.g., 5 mV/s) point to the characteristics of quasi microelectrode, suggesting that the ratio of interelectrode distance to diameter of 14 is still not sufficiently large, and that slight overlap of the diffusion layers (shield effect) begin to appear. Similar features were observed also by using choronoamperometry. At a timescale of 1000 s, the steady or quasi-steady state is dominant with the UMEAs, which could be confirmed by the relative constant current response.

Ideally, microelectrode arrays should not show any overlap of diffusion layers or any shield effect. Identical sigmoidal cyclic voltammograms for a sufficient long times and a constant chronoamperometric response (limiting current) at steady state should be observed. However, in practice, in order to increase the current signal and to reduce the size of the chip and the cost of fabrication, arrays with a relative high packing density of microelectrodes are usual, though this is realized at the expense of the occurrence of the shield effect. To some extent, arrays with some overlap of diffusion layers can be tolerated, provided the steady state current is not seriously affected. As a matter of fact, for the design of ultramicroelectrode arrays, perfection is neither necessary nor expected. The slight loss of independence of microelectrodes should not be a problem in the SWASV technique, because the experiment is usually performed under flowing mode (stirring) by pumping. Nevertheless, in principle microelectrode arrays with a larger interelectrode distance and smaller dimensions are preferred (WANG 2000).

6. Analytical performance of ultramicroelectrode arrays

6.1 Materials and methods

6.1.1 Apparatuses and reagents

Chronocoulometry, square wave anodic stripping voltammetry, and linear sweep voltammetry were carried out using an Autolab PGSTAT 10 potentiostat (Eco Chemie, Netherlands). Pictures of mercury film were taken by means of optical microscope (Zeiss, Germany) equipped with software of digital image analyser (DIANA) and a Sanyo colour CCD digital camera. Surface morphology of UMEs was recorded using atomic force microscope (AFM) (Dimension 3100, Digital Instruments, USA) with NanoScope IV Control software. Energy dispersive X-ray analysis (EDX) was performed by means of a Camebax S x 50 microprobe (CAMECA, France). UMEA chips were cleaned in ultrasonic bath (Sonorex, RK 100, Germany). Solutions of trace concentrations of heavy metals (Cu, Pb, Cd, and Zn) were prepared by 1000 mg/l AAS standards (Alfa, Germany). Mercury film was plated in a solution of 5 x 10^{-4} M mercury (II) nitrate monohydrate (Merk, analytical grade) in 0.1 M nitric acid (Merk, Suprapur). As supporting electrolyte solution, 0.02 M acetate buffer was prepared with anhydrous sodium acetate (Merck, 99.99%, Suprapur) and glacial acetic acid (Merck, 100% Suprapur). Mercury film was oxidized with a solution of 1 M potassium thiocyanate (Merk, analytical grade). All solutions were prepared with double distilled or milli-Q water.

6.1.2 Electrochemical methods

After characterization of UMEAs, mercury film was generated using chronocoulometry under a constant plating potential of -0.3 V in a 5 x 10^{-4} M $Hg(NO_3)_2$/0.1M HNO_3 solution and its quality was inspected by optical microscope. The analytical performance of UMEAs was investigated by determination of trace concentrations of Cu, Pb, Cd, and Zn in 0.02 M acetate buffer solutions using square wave anodic stripping voltammetry (SWASV). Mercury film was oxidized in a solution of 1 M potassium thiocyanate using linear sweep voltammetry in a potential range from -0.2 V to 0.2 V and with a scan rate of 10 mV/s.

6.2 Results and discussion

6.2.1 Optimum conditions for generating mercury film

Mercury film quality plays an important role in the trace metal measurements using UMEAs. The prerequisite for successful measurements is an optimum mercury film. A lot of parameters have influence on the formation of the mercury film, including plating potential, mercury concentration of plating solution, plating time or charge, pump rate, electrode material etc.. Under different plating conditions, planar, hemispherical, or spherical mercury film can be observed. However, considering the sensitivity of measurement and the stability of mercury film, hemispherical mercury film proves to be the most suitable form for the stripping voltammetry (Fig. 24).

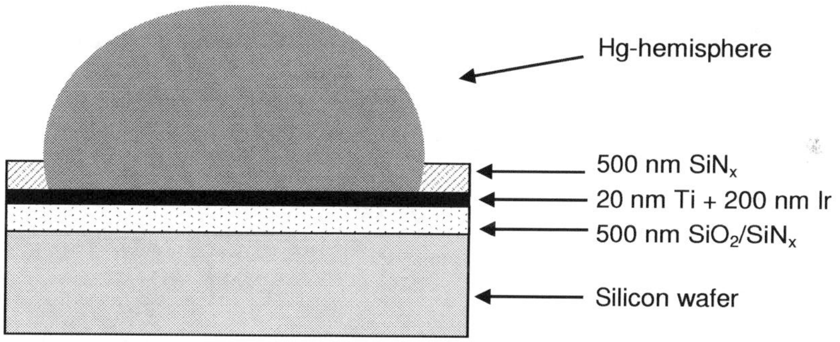

Fig. 24: Schematic representation of hemispherical mercury film formed on an Ir microelectrode

In this study, because of the very small size of UMEs, a less negative plating potential of -0.3 V, a low mercury concentration of 5 x 10^{-4} M $Hg(NO_3)_2$, and a pumping rate of 0.33 ml/min was used for plating. Instead of chronoamperometry, chronocoulometry (charge vs. time) was applied for plating, due to the fact that charge is a better criterion to define the duration of the plating in getting an optimum mercury film (SILVA ET AL. 1999). Figures 25a and 25b show the relationship between plating times and plating charges using chronocoulometry, demonstrating different plating rates (mC/s) using the same chip.

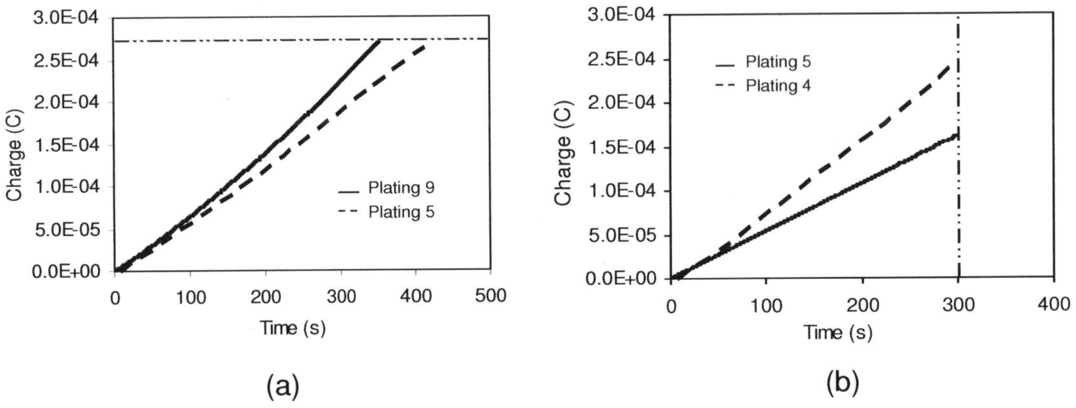

Fig. 25: Plating of mercury film by means of chronocoulometry (a) different plating times and constant charge (0.27 mC) for chip 1, (b) different charges and constant plating time (300 s) for chip 2

Experiments show that plating rate decreases after several measurements, which is possibly as a result of the fouling of the chip. Therefore, Hg film of good quality can be only achieved, if an optimum plating charge, instead of plating time, was determined (TERCIER ET AL. 1995). For four arrays, the optimum charge was found to be in the range of 0.20 ~ 0.30 mC. If the charge was less than 0.20 mC, the sensitivity of the measurements decreased substantially; if the charge was higher than 0.30 mC, large Hg drops were formed, which were unstable and tended to remove from electrodes to the passivation layer. Mercury film quality was examined by optical microscope during the plating process. As examples, figure 26 shows the mercury film photos of a chip obtained with optical microscope at different plating charges.

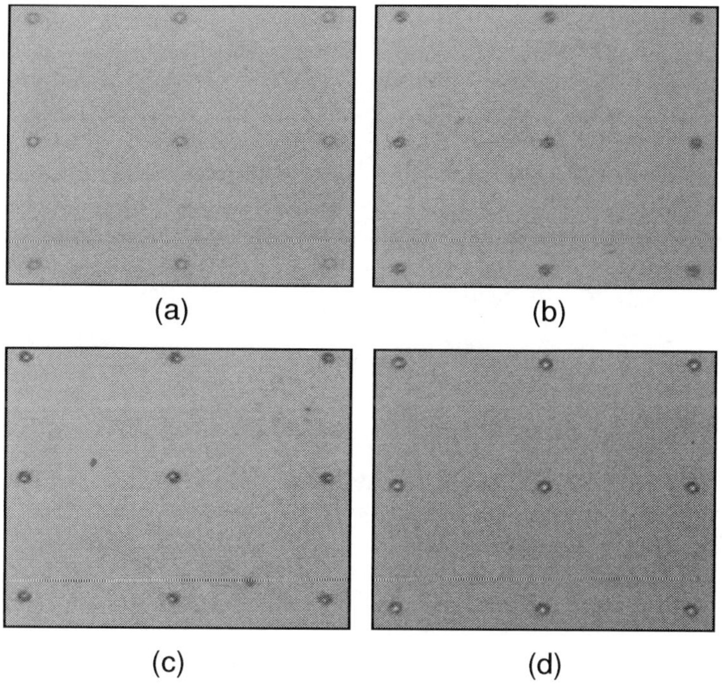

Fig. 26: Examples of optical microscope photos during the plating of mercury film (a) new chip, (b) 0.15 mC for a plating time of 159 s, (c) 0.20 mC for a plating time of 260 s, (d) 0.27 mC for a plating time of 353 s

6.2.2 Analytical performance of UMEAs

After the generation of an optimum Hg film, trace concentrations of Cd, Pb, and Cu, in synthetic aqueous solutions were determined, using square wave anodic stripping voltammetry (SWASV). Sensitivity, precision, and accuracy were tested with different Ir chips and different Hg films.

6.2.2.1 Sensitivity

There are mainly three parameters which significantly influence the measurement sensitivity: mercury film quality, deposition time, and square wave frequency. For an optimum Hg film, the sensitivity is proportional to deposition time and frequency (WANG 1985). In this study, sensitivities as low as 0.25 µg/l of Cd and Pb could be achieved using a deposition time of 360 s and a frequency of 300 Hz, which corresponds to 175 nA/µg/l for Cd and 205 nA/µg/l for Pb. For the measurement of

1.00 µg/l Cd and Pb, the deposition time was reduced to 240 s, which corresponds to 108 nA/ug/l and 131 nA/µg/l, respectively (Fig. 27).

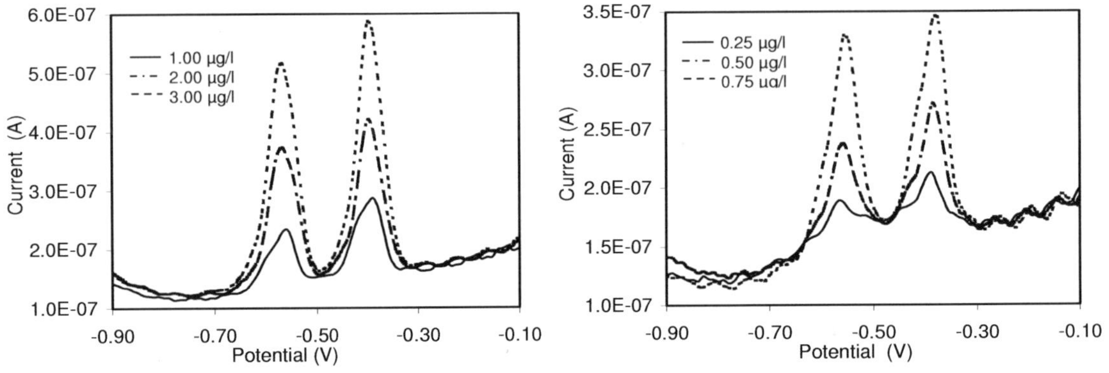

(a) (b)

Fig. 27: Typical SWASV voltammograms obtained with an Ir-UMEA chip in a concentration range of 0.25 - 3.00 µg/l Pb and Cd in a 0.02 M acetate buffer solution (pH = 4.5). Analytical conditions were: deposition potential, -1.2 V; frequency, 300 Hz; amplitude, 20 mV; step height, 10 mV. (a) 1.00, 2.00, 3.00 µg/l Pb and Cd, deposition time, 240s, (b) 0.25, 0.5, 0.75 µg/l Pb and Cd, deposition time, 360 s

6.2.2.2 Precision

Reproducibility was estimated by repetitive measurements of 1.00 µg/l Pb and Cd synthetic standard solutions. As examples, a summary of the results obtained with two chips (Ir 24 and Ir 27) is given in table 2 and in appendix 1.

Tab. 2: Repetitive measurements of 1.00 µg/l Cd and Pb in 0.02 M acetate aqueous solutions

Chip	Plating	Element	C (µg/l)	N	8	SD	RSD (%)
Ir 27	Plating 5	Cd	1.00	28	0.97	0.11	11
		Pb	1.00	20	1.06	0.13	13
	Plating 9	Cd	1.00	50	0.94	0.11	11
		Pb	1.00	50	1.00	0.09	9
Ir 24	Plating 4	Cd	1.00	45	1.00	0.11	11
		Pb	1.00	42	1.01	0.12	12

C, concentration; N, number of measurements; 8, mean value; SD, standard deviation; RSD (%), relative standard deviation

Based on 123 measurements of Cd, a standard deviation of 0.11 was found, corresponding to a relative standard deviation of 11%; the standard deviations of 112 measurements of Pb for different platings were in the range of 0.09 ~ 0.13, corresponding to relative standard deviations of 9 ~ 13%. As an example, 50 repetitive measurements were performed using only one mercury film (plating 9 of Ir 27) are schematically presented in figure 28.

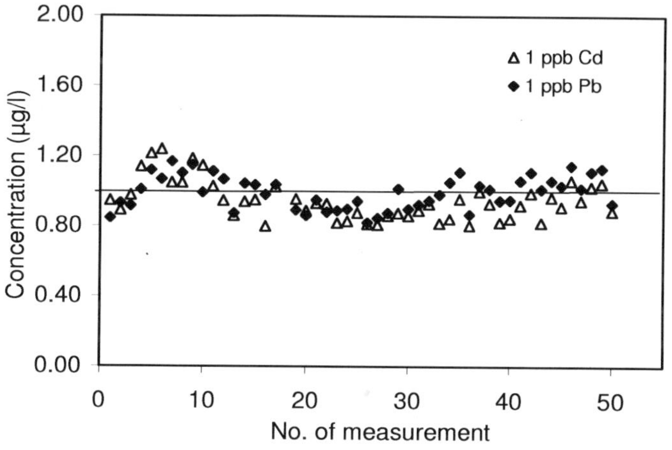

Fig. 28: Repetitive measurements of 1.00 µg/l Cd and Pb in 0.02 M acetate buffer solution (pH = 4.5), using a single mercury film. The SWASV conditions were: deposition potential –1.2 V, depostion time 240 s, frequency 300 Hz, amplitude 25 mV, and step potential 15 mV

6.2.2.3 *Accuracy*

Accuracy was evaluated by repetitive measurements of the reference sample "Trace Metals in Drinking Waters" (High Purity Standards, USA) (Tab. 3 and appendix 1).

Tab. 3: Measurements of the certificated reference sample (Trace Metals in Drinking Water, HPS, USA); dilution factor 10 for Cd and 40 for Pb

Chip	Plating	Element	C (µg/l)	N	8	SD	RSD (%)
Ir 29	Plating 1	Cd	1.00	9	0.94	0.10	10
		Pb	1.00	9	0.91	0.12	12
	Plating 2	Cd	1.00	5	1.04	0.06	6
		Pb	1.00	5	1.09	0.09	9

C, concentration; N, number of measurements; 8, mean value; SD, standard deviation; RSD (%), relative standard deviation

Concentrations of 1.00 µg/l Cd and Pb were prepared by diluting the standard (diluting factor 10 and 40, respectively). Based on two different platings, the average measurement values are within ± 6% and ± 9% of the certified value of Cd and Pb, respectively.

6.2.2.4 Comparison of measurements with a single array and all four arrays interconnected

Results of measurements obtained with a single array and with four arrays together are compared in figure 29. The current intensity measured with four arrays are approximately four times higher than with a single array, demonstrating a collective current response of array electrodes. The significantly enhanced current signal is advantageous for measurements using conventional electrochemical apparatuses, especially for very low concentration levels of heavy metals (below µg/l). However, it should be mentioned that the background currents using four interconnected arrays increase also at the same time in comparison with a single array. This means that by using four arrays simultaneously the signal to noise ratio is not necessarily improved and thus the sensitivity is not necessarily increased. Nevertheless, the lifetime of a chip can be substantially prolonged with the array-design, which is important for the development of in-situ and on site instrument for the determination of trace concentrations of heavy metals in natural waters.

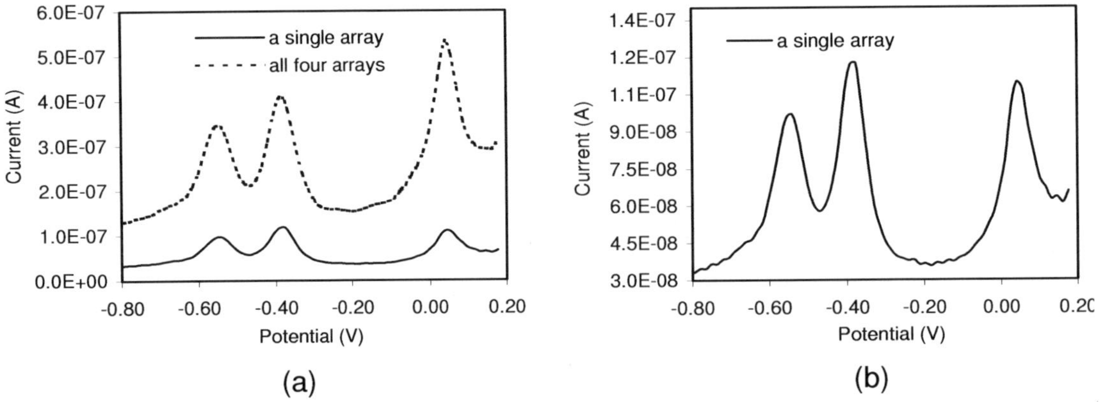

Fig. 29: Comparison of voltammograms obtained with the measurements of 1.00 µg/l Cu, Pb, and Cd in 0.02 M acetate buffer solutions (pH = 4.5): (a) a single array and all four arrays together, (b) a single array. The SWASV conditions were: deposition potential, –1.2 V; depostion time, 180 s; frequency, 300 Hz: amplitude, 25 mV; and step potential, 15 mV

6.2.3 Oxidation of mercury film – deplating

After several measurements, the quality of mercury film gets degraded and, the sensitivity of the SWASV measurements get worse, indicating the necessity of the generation of a new mercury film. Prior to new plating, rests of old mercury film must be removed. Generally, two kinds of methods are used for the oxidation of the mercury film: electrochemical- and chemical methods. By electrochemical method, the mercury film is oxidized in a solution of 1 M potassium thiocyanate by linear sweep voltammetry in the potential range of -0.2 ~ 0.2 V (DIEDERICH ET AL. 1994). The oxidation of mercury is manifested by an oxidation peak occurring at a potential of about 0 V. Incidentally, the oxidation of mercury film yields a lower charge as compared to the charge needed for the generation of mercury film (SILVA ET AL. 1999, BELMONT 1996), which is difficult to explain. As for the chemical method used in this study, the mercury film is simply oxidized by concentrated acids (e.g., HNO_3) (KOUNAVES ET AL. 1994, SILVA ET AL. 1999), which is however possible only in laboratory and is not feasible for in situ measurements.

6.2.4 Lifetime of UMEAs

It was found that an Ir-UMEA chip with good quality could sustain the plating/deplating operation of mercury film at least 10 times, allowing about 500 measurements of trace metal concentrations in synthetic aqueous solutions. After that it was difficult to get an optimum mercury film. The surface roughness and the chemical composition of the UMEs were investigated by means of AFM and EDX analysis, respectively, in order to get an insight into the parameters which influence the lifetime of the chips.

6.2.4.1 Roughness analysis of UME surfaces

A comparison of the surface roughness of UMEs of a new and a used UMEA chips is shown in figures 30a and 30b. The roughness index of UME surface obtained with AFM changes from 1.0 nm of the new chip to 8.3 nm of the used one. This means that after some time, the surfaces of the electrodes are covered by dirty accumulations, which could alter the size of UMEs or even block the UME surfaces, preventing the UMEs to work properly. The accumulations may be caused by small particles from flow through cell (e.g., the rubber of O-ring), from pumping tube, or from by-products of chemical reactions. It was reported that calomel could be formed in chloride rich media and accumulated on the surface of UMEs. The contamination of chloride in synthetic solutions resulted mainly from leakage of reference electrode, or from the use of Cl^--containing chemicals like $HClO_4$, NaCl, HCl etc. (NOLAN & KOUNAVES 1998). Reactions of the chip compositions (e.g., passivation materials and photoresist) with chemicals used in experiments or breakdown of $K_3Fe(CN)_6$ during the cyclic voltammetryic tests may also cause the accumulations of dirt, and consequently shorten the lifetime of the chip (SCHMITT ET AL. 1999, FASSBENDER ET AL. 2000, SCHMITT ET AL. 2000) .

For cleaning (or activation) of solid electrodes, several methods are suggested, including mechanical polishing, chemical or electrochemical treatment, and physical modification by heat, plasma, or laser radiation (STULIK 1992). However, only electrochemical and chemical methods are commonly used for microelectrodes, because the structures of microelectrodes tend to be very thin and are mostly in a recessed form (WIGHTMAN 1998, WANG ET AL. 1996), making the resurfacing of microelectrodes by mechanical polishing impossible (FIACCABRINO & KOUDELKA-HEP

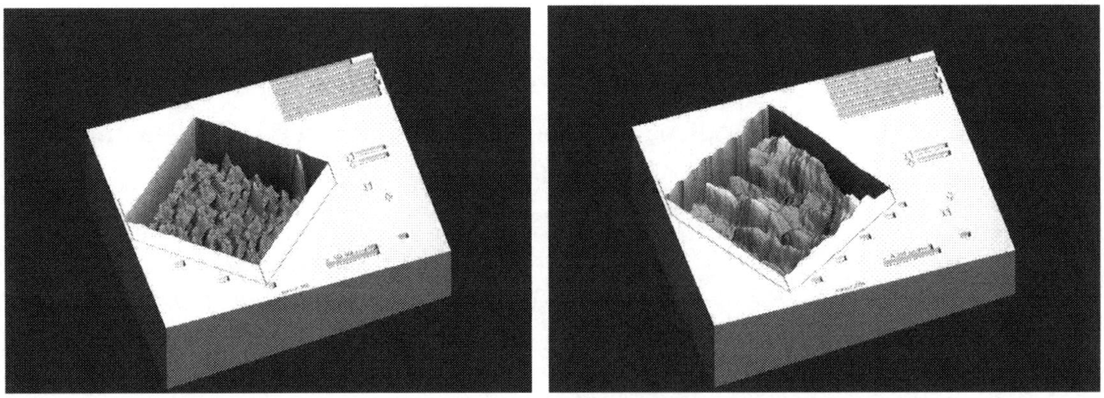

(a) (b)

Fig. 30: Comparison of the electrode surface roughness of a new and a used Ir-UMEA chips by AFM: (a) the new chip with an electrode roughness index of 1.0 nm, (b) the used chip with an electrode roughness index of 8.3 nm

1998). In this study, the electrode surface of UMEs was cleaned either electrochemically by generating hydrogen microbubles at a potential of –1.9 V in a 0.05 M HNO_3 solution, or chemically by use of concentrated HNO_3 and/or H_2SO_4 or organic solvents (e.g., CH_3COCH_3, CH_3OH). Unfortunately, both of the methods proved to be ineffective. Cleaning in ultrasonic bath has proved to be a very effective method to get rid of such dirty accumulations. However, an excessive use of ultrasonic bath may damage the bond wires of chip.

It should be noted that, in order to avoid the formation of calomel, no Cl^--containing solutions were used and the micro-flow-through reference electrode (Ag/AgCl, 3 M KCl) was placed downstream of the chip in the process flow during the experiments. By a thorough cleaning of the O-ring and the micro-flow-through cell by means of ultrasonic bath, the fouling caused by plastic particles could be successfully eliminated.

6.2.4.2 EDX analysis

An Ir/Pt chip, which stopped to work properly in experiment, was investigated by means of EDX. The chemical composition spectra of the electrode and passivation layer are shown in figure. 31. The absence of Ir peak at electrode indicates the

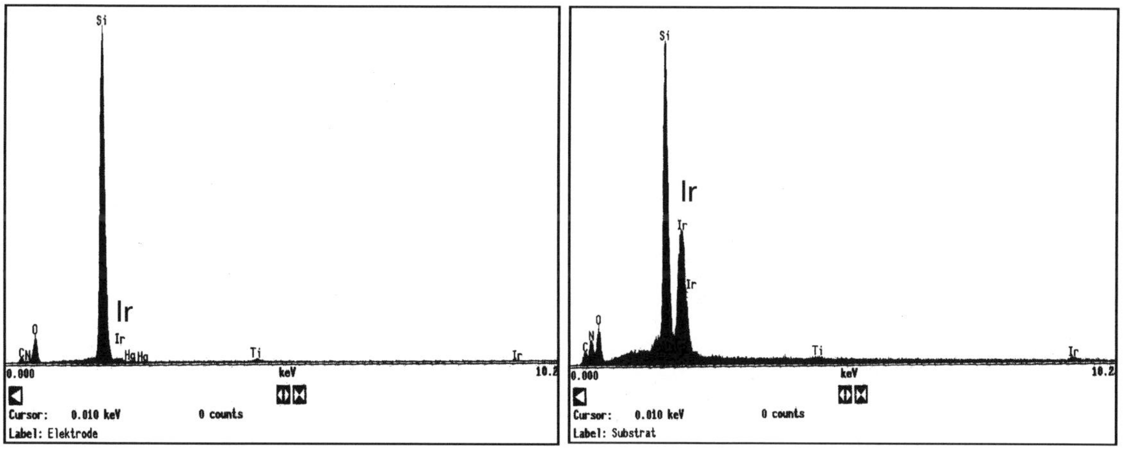

(a) (b)

Fig. 31: Comparison of the EDX spectra of a used Ir/Pt chip (a) electrode, and (b) passivation layer removal of Ir from UME (Fig. 31a). In contrast, as shown in figure 31b still a large Ir peak is observed in the passivation layer. This demonstrates that Ir was worn out from the exposed UMEs after a long time use of the chip. The exact reason for the removal of Ir is not clear, but it is possibly due to the poor fabrication quality of the Ir/Pt chip or to an inappropriate chemical or electrochemical treatment of the chip during the experiments. However, under almost identical measurement conditions, such problem did not find with Ir-UMEA chips, demonstrating a good fabrication quality of the Ir chips.

7. Optimisation of experimental parameters

The coupling of ultramicroelectrode arrays with square wave anodic stripping voltammetry (SWASV) has been proved to be a promising method for the determination of trace concentrations of heavy metals in water. However, for a successful application of this method, a lot of factors have to be considered, including instrumental parameters, experimental conditions, and interferences. In order to benefit from the advantages of the square wave technique (SW), parameters such as frequency, step potential, amplitude, deposition potential, and deposition time were optimised. Since interferences may become problematic in SWASV measurements, the influences of the most important parameters, such as dissolved oxygen, intermetallic compounds, dissolved organic carbon (DOC), and chloride ions, were investigated. The effects of experimental conditions, like the composition of the supporting electrolyte, pH value, and medium exchange, etc. were also studied.

7.1 Selection of instrumental parameters

7.1.1 Frequency of SW

The effect of SW frequency on the SWASV measurement is illustrated in figure 32. The integrated peak current intensity increases proportionally with increasing square wave frequency up to a value of 400 Hz. For a frequency higher than 400 Hz, the peak currents nearly level off (Fig. 32a). The voltammogram in figure 32b shows that a higher square wave frequency yields a higher sensitivity of the ASV measurement but at the expense of poorer peak resolution, worse peak baseline, and broader peak shape. The optimum frequency range is between 200 and 400 Hz, in which an obvious improvement in both peak current and peak resolution is obtained. In addition, the voltammograms in figure 32b show also that the peak potentials shift slightly towards positive values with increasing frequency. Compared to 200 Hz, a shift of 45 mV in peak potential is observed at a frequency of 700 Hz.

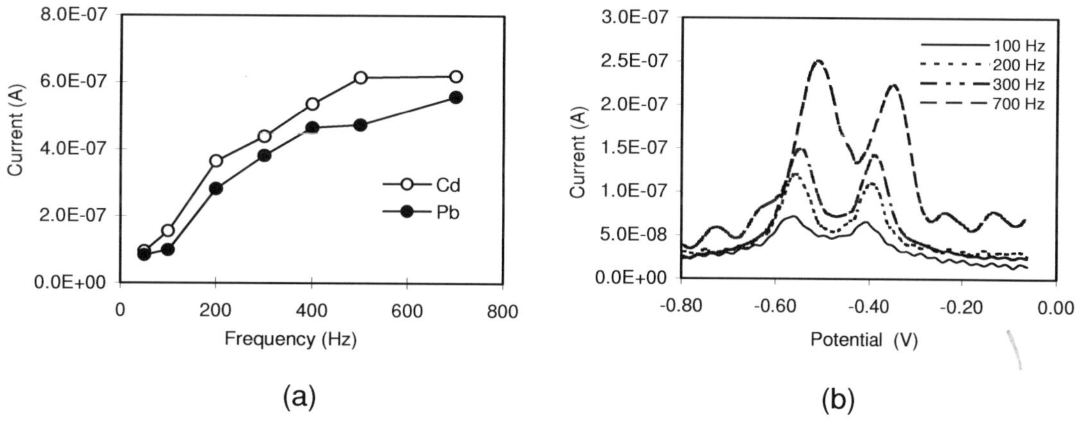

Fig. 32: Effect of SW frequency on the SWASV measurement of 1.00 µg/l Cd and Pb in a 0.02 M acetate buffer solution (pH = 4.5). Effect on (a) peak current intensity, and (b) voltammograms. Analytical conditions: deposition potential, -1.2 V; deposition time, 180 s; amplitude, 25 mV; step potential, 15 mV

7.1.2 Step potential of SW

The effect of SW step potential on the SWASV measurements was studied by varying the step potential from 5 mV to 30 mV. As shown in figure 33a, an obvious increase of sensitivity occurs at step potentials lower than 15 mV. In the range of 15 mV and 30 mV, the peak current intensity increased slowly. The examples of

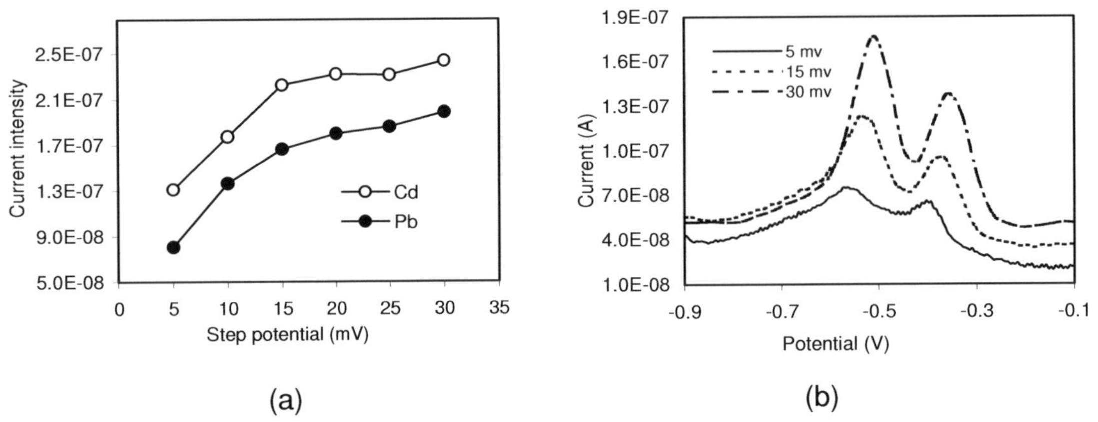

Fig. 33: Effect of SW step potential on the SWASV measurement of 1.00 µg/l Cd and Pb in 0.02 M acetate buffer solution (pH = 4.5). Effect on (a) peak current intensity, (b) voltammograms. Analytical conditions: deposition potential, -1.2 V; deposition time, 180 s; square wave frequency, 300 Hz; amplitude, 25 mV

voltammograms in figure 33b show that no improvement in peak resolution and peak baseline is obtained by changing the step potential. Larger background currents and a slight shift of the peak potentials towards positive values are observed by increasing the step potential. The optimum step potentials range from 10 mV to 20 mV.

7.1.3 Amplitude of SW

The effect of SW amplitude on SWASV measurements was examined by varying the SW amplitude from 10 mV to 40 mV. As shown in figure 34a, a proportional increase in current response was found in the amplitude range of 10 mV and 25 mV. Figure 34b indicates that with increasing SW amplitude, the baseline of the voltammogram shifts gradually towards higher currents. No influence was observed on the position of peak potential and peak resolution, though a smoother baseline was found at lower SW amplitudes. Considering the sensitivity, the optimum SW amplitudes should be in the range of 15 mV and 25 mV.

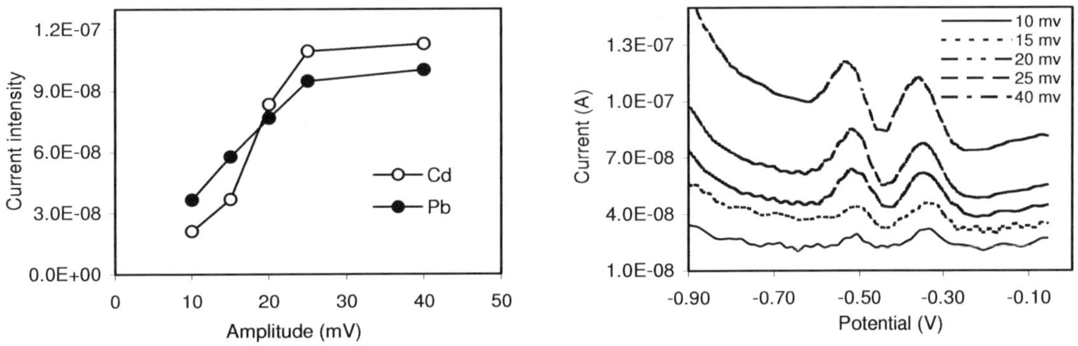

(a) (b)

Fig. 34: Effect of SW amplitude on the SWASV measurement of 1.00 µg/l Cd and Pb in 0.02 M acetate buffer solution (pH = 4.5). Effect on (a) peak current intensity, (b) voltammograms. Analytical conditions: deposition potential, -1.2 V; deposition time, 180 s; square wave frequency, 300 Hz; step potential, 15 mV

7.1.4 Deposition potential

Because the application of ASV is limited by the reduction of hydrogen ions and the oxidation of mercury, the deposition potential is usually applied in the range of 0 and -1.5 V for the determination of heavy metals in aqueous solutions (VAN DEN BERG 1989). In the preconcentration step, the deposition potential is usually 0.3 ~ 0.5 V more negative than the half-wave potential ($E_{1/2}$) of the least reduced metal ions (WANG 2000). Because the half-wave potentials of Cd and Pb in 0.02 M acetate buffer solution are approx 0.54 V and 0.39 V (vs. Ag/AgCl reference electrode), respectively, a potential in the range of -0.8 and -1.4 V was chosen for SWASV measurement. The effect of deposition potential on the measurements of 1.00 µg/l Cd and Pb in 0.02 M acetate buffer is shown in figure 35. It can be seen from figure 35a that the peak current increases proportionally with deposition potential until -1.2 V and then nearly levels off. The voltammograms in figure 35b shows that in the potential range of -1.2 V and -1.4 V the sensitivity of measurement increased substantially, together with the improvement in the peak shape. Moreover, a very slight shift of the peak potential towards negative values was observed, if a more negative deposition potential was applied. Consequently, the selection of a more negative potential is advantageous, if the interference is not severe during the SWASV measurement. For simultaneous determination of trace concentrations of Pb and Cd in 0.02 M acetate buffer solution (pH = 4.5), the optimum potentials lie in the range of -1.2 V and -1.4 V.

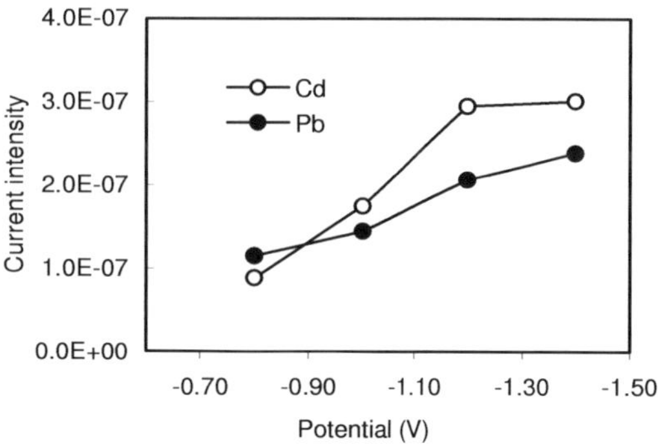

(a)

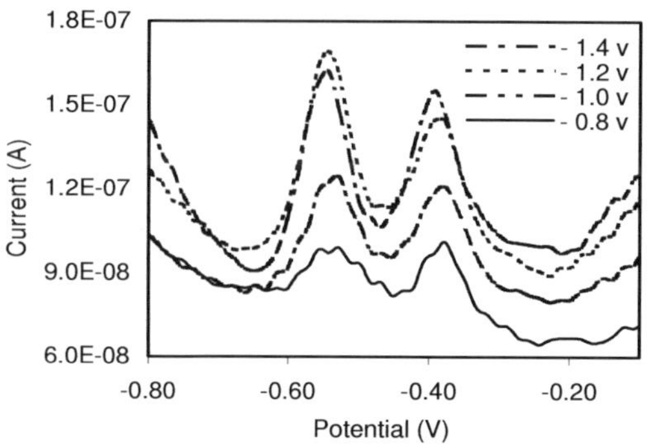

(b)

Fig. 35: Effect of deposition potential on the SWASV measurement of 1.00 µg/l Cd and Pb in 0.02 M acetate buffer solution (pH = 4.5). Effect on (a) peak current intensity, (b) voltammograms. Analytical conditions: deposition time, 180 s; square wave frequency, 300 Hz; amplitude, 25 mV; step potential 15 mV

7.1.5 Deposition time

Deposition time plays a very important role in ASV measurement. The extremely low detection limits (up to parts per billion) achieved by ASV are attributed mainly to the long deposition times. The dependence of peak currents on deposition time in the measurement of 1.00 µg/l Cd and Pb in 0.02 M buffer solution is shown in figure 36. In the time range of 60 and 360 s, the peak current signals increase almost linearly with time. However, it is known that too long deposition times may lead to the overloading (or saturation) of the electrodes with metal ions and to formation of intermetallic compounds (METROHM 1994, SOTTERY & ANDERSON 1987), especially at high concentration levels. Therefore, depending on the concentration of metal ions, a suitable deposition time should be applied. Nevertheless, for the measurement of Cd and Pb, no such saturation was observed in this study, even at higher concentration levels (e.g., 10 µg/l) and longer deposition times (300 s).

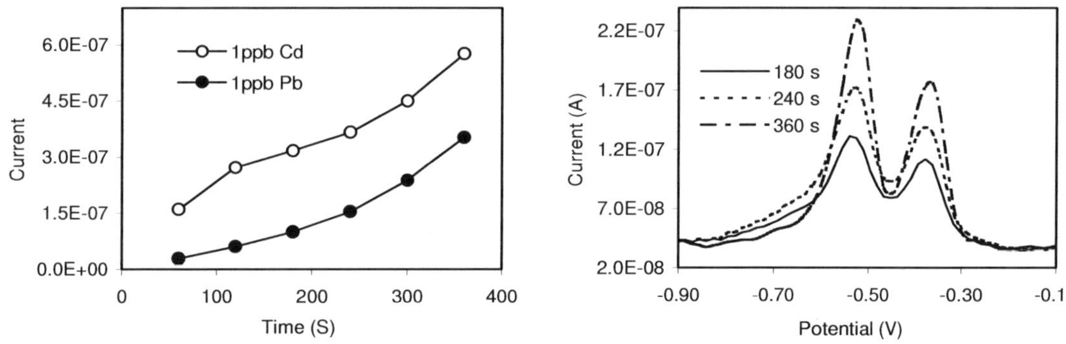

(a) (b)

Fig. 36: Effect of deposition time on the SWASV measurement of 1.00 µg/l Cd and Pb in 0.02 M acetate buffer solution (pH = 4.5). Effect on (a) peak current intensity, (b) voltammograms. Analytical conditions: deposition potential, -1.2 V; square wave frequency, 300 Hz; amplitude, 25 mV; step potential 15 mV

7.2 Selection of experimental conditions

7.2.1 Supporting electrolyte

In order to minimize the ohmic drop effect, an inert supporting electrolyte is needed in anodic stripping voltammetry. Depending on the application, the electrolyte can be an inorganic salt (e.g., KCl, KNO_3), a mineral acid, (e.g., HCl, HNO_3, $HClO_4$), a buffer solution (e.g., acetate-, citrate buffer), if pH control is needed, or even a basic solution. The composition of the supporting electrolyte plays an important role in the deposition potential applied on the working electrodes, in the position of peak potential (i.e., selectivity), and the intensity of peak current (i.e., sensitivity). A suitable electrolyte could lead to the improvement of voltammograms, such as well-defined peaks, no peak overlapping, smooth peak baselines, etc.. For conventional electrodes, the concentration of the supporting electrolyte is usually about 0.1 M (WANG 1985), which is an compromise between high conductivity and minimum contamination. For microelectrodes, the concentration of electrolyte can be much lower (e.g., 0.01 M) or even no electrolyte (CISZKOWSKA & STOJEK 2000), due to the immunity of microelectrodes to ohmic drop. Consequently, As a result, the contamination caused by addition of electrolyte is highly minimized.

The effects of different electrolytes on the ASV measurement of 1.00 µg/l Cd, Pb, and Cu in aqueous solution were studied. As electrolytes 0.02 M acetate buffer (pH = 4.5), 0.1 M $NaNO_3$, 0.02 M acetate buffer+0.05 M NaNO3, and 0.1 M HNO_3 were tested. Figures 37a and 37b show the results of the measurements in 0.02 M acetate buffer and in 0.1 M $NaNO_3$. Compared to 0.02 M acetate buffer, the peak currents

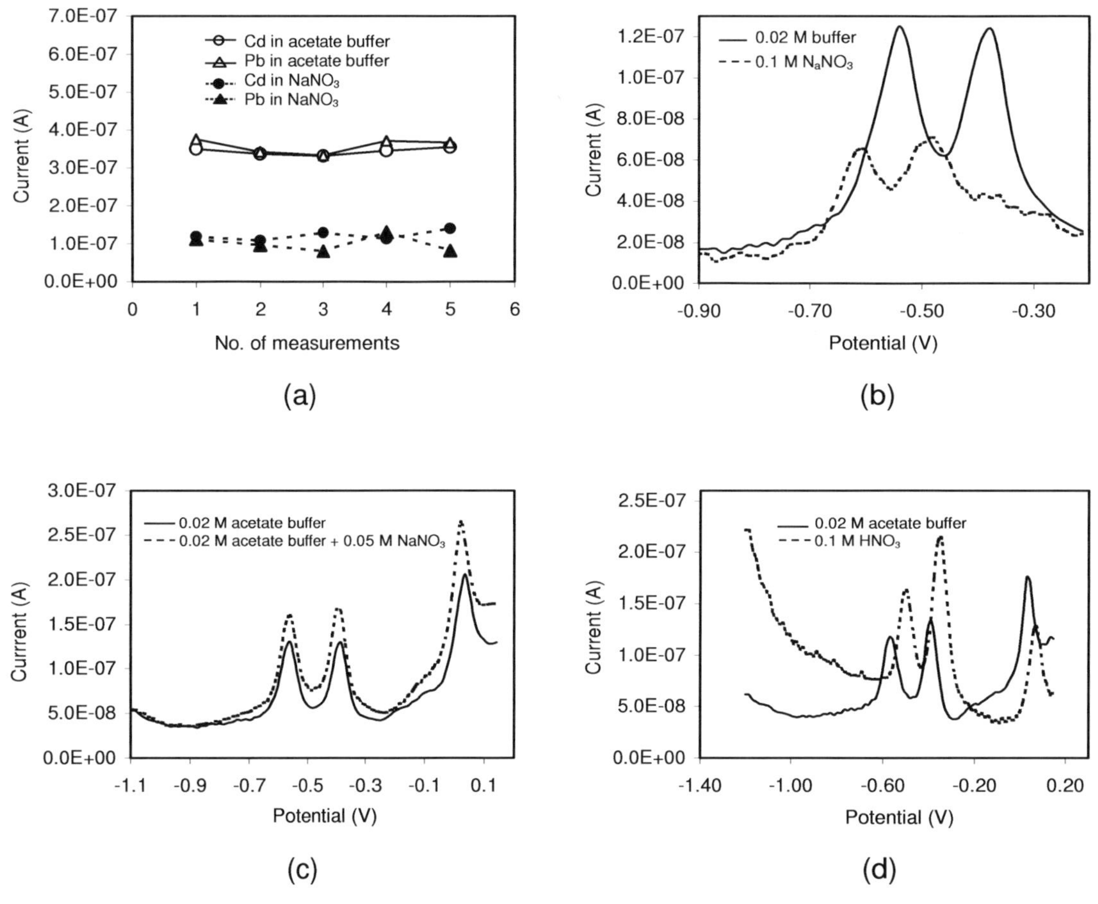

Fig. 37: Comparison of the SWASV measurements using different supporting electrolytes. (a) peak currents of 1.00 µg/l Cd and Pb in 0.02 M acetate buffer (pH = 4.5) and in 0.1 M $NaNO_3$, (b) voltammograms of 1.00 µg/l Cd and Pb in 0.02 M acetate buffer (pH = 4.5) and in 0.1 M $NaNO_3$, (c) voltammograms of 1.00 µg/l Cu, Pb, and Cd in 0.02 M acetate buffer (pH = 4.5) and in 0.02 M acetate buffer+0.05 M $NaNO_3$, (d) voltammograms of 1.00 µg/l Cu, Pb, and Cd in 0.02 M acetate buffer (pH = 4.5) and in 0.1 M HNO_3. Analytical conditions: deposition potential, -1.2 V; square wave frequency, 300 Hz; amplitude, 25 mV; step potential 15 mV

measured in 0.1 M $NaNO_3$ are much lower, indicating lower sensitivity. As shown in the voltammograms, a slight shift of peak potential towards negative values occurs in 0.1 M $NaNO_3$, which may be caused by the change of pH.

A comparison of the measurements in 0.02 M acetate buffer (pH = 4.5) and in 0.02 M acetate buffer + 0.05 M $NaNO_3$ is shown in figure 37c. In the 0.02 M acetate buffer + 0.05 M $NaNO_3$ solution, no improvement in sensitivity is observed. On the contrary, a higher background current and a worse peak baseline result. Consequently, for SWASV measurements using UMEA sensor, it is of no advantage to strengthen the supporting electrolyte. This proves that microelectrodes can be utilized in solutions with relative high resistance, in which a larger ohmic drop (iR) will be developed with conventional electrodes.

Figure 37d shows the SWASV voltammograms of Cu, Pb, and Cd recorded in solutions of 0.1 M HNO_3 and 0.02 M acetate buffer. In both electrolytes, satisfactory voltammograms were obtained. However, for the measurement in 0.1 M HNO_3 solution, a larger background current is observed, if the deposition potential gets more negative (e.g., < -1.2 V), due to the reduction of hydrogen ions. The large background current hampers seriously the determination of trace concentration of Zn. It should be mentioned that the mercury film got quickly damaged in acidic electrolyte, which is probably caused by the hydrogen gas bubbles evolved on the surface of the UMEA chip and by the chemical reaction between mercury and nitric acid.

In conclusion, 0.02 M acetate buffer (pH = 4.5) prove to be the optimal supporting electrolyte for the ASV measurement of trace concentrations of heavy metals in aqueous solutions using UMEA sensor. With this electrolyte, a relative high sensitivity could be attained and at the same time the mercury film could be used for a long time. Though at potentials more positive than -1.2 V a 0.01 M HNO_3 solution is also a possible alternative for the measurements, the lifetime of the mercury film gets substantially shortened.

7.2.2 pH

The influence of different pH values on the sensitivity of the ASV measurement in 0.02 M acetate buffer is illustrated in figure 38. As shown in figure 38a the peak heights of 1.00 µg/l Cd and Pb increase as the pH values get higher than 4, especially for Cd. In contrast, at pH value between 3 and 4, the influence of pH on the measurement is less evident and the sensitivity is much lower. Therefore, for the measurement of Cd and Pb, the optimum pH should be higher than 4. The influence of pH on the SWASV voltammograms is shown in figure 38b. Apart from the peak current intensity, the influence of pH on the peak potential of Cd and Pb is also shown. It is evident that the lower the pH values are, the more positive are the peak potentials of Cd and Pb. In general, for the simultaneous determination of different heavy metals (e.g., Cu, Pb, Cd, and Zn) in aqueous solutions, too high or too low pH values should be avoided, because (i) a high pH may cause the precipitation of heavy metal hydroxides, and (ii) a low pH facilitates the reduction of hydrogen ions to hydrogen gas, which is disadvantageous for measurements, especially for Zn. In acid solutions the determination of Zn suffers from the interference of hydrogen wave, because the reduction potential of Zn at the mercury electrodes is more negative than, or is equal to the potential of the reduction of hydrogen ions (CROSMUN ET AL. 1975). In principle, the reduction of hydrogen ions in acidic media follows the reaction is as follows:

$$2\,H^+ + 2\,e^- \rightarrow H_2$$

According to the Nernst equation, the relationship between pH and the reduction potential of hydrogen ions can be expressed as:

$$E = E^0_{H_2/H^+} - 0.059\ pH$$

Therefore, one unit change of pH will cause a potential shift of 59 mV. The more acidic the solution, the more positive is the reduction potential of the hydrogen ions.

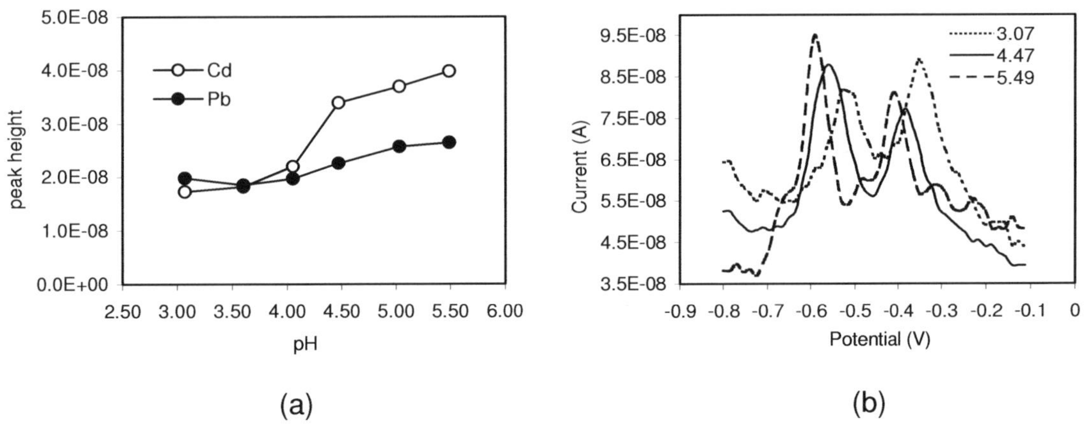

Fig. 38: Influence of pH on the SWASV measurement of 1.00 µg/l Cd and Pb in 0.02 M acetate buffer solutions. Effect on (a) peak current intensity, (b) voltammograms. Analytical conditions: deposition potential, -1.2 V; square wave frequency, 300 Hz; amplitude, 25 mV; step potential 15 mV

7.2.3 Medium exchange

Medium exchange is an effective method for the improvement of the ASV measurement of heavy metals in complex media (DESLMONI ET AL. 1980, WAHDAT & NEEB 1987). The method involves the deposition of heavy metals in sample solution, followed by the striping of heavy metals in a more suitable electrolyte (WANG & GREENE 1983). By changing to a pure electrolyte, the electrodes are completely washed, leading to an increase in the selectivity of measurement. Problems (e.g., peak overlapping, interferences due to organic matter and dissolved oxygen) could be substantially reduced (WANG & GREENE 1983, WAHDAT & NEEB 1987).

In this study, the river water samples were not UV-digested and only buffered to a 0.02 M acetate solution. After the deposition step, the stripping of heavy metals was performed in a pure deaerated 0.02 M acetate buffer solution with a pH value of approx 4.5. The results of the measurements with and without exchanging electrolyte are compared in figure 39. An obvious improvement in both the shape of voltammograms (higher resolution) and the sensitivity of measurements is indicated after electrolyte exchange.

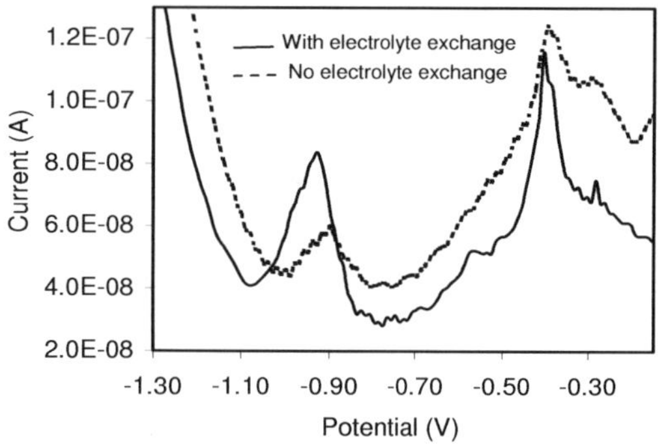

(a)

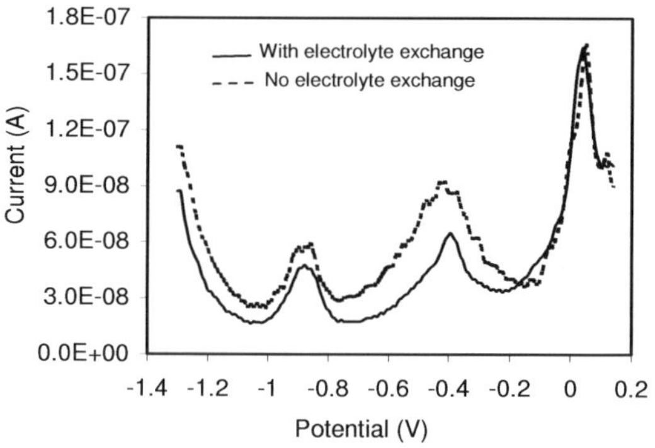

(b)

Fig. 39: Comparison of SWASV measurements of trace concentrations of heavy metals in river water with and without electrolyte exchange. Analytical conditions: deposition potential, -1.4 V; deposition time, 240s; square wave frequency, 300 Hz; amplitude, 25 mV; step potential 15 mV

7.3 Interferences

7.3.1 Dissolved oxygen

It is well known that dissolved oxygen interferes with the analyte signals in ASV using conventional electrodes, and has to be removed by purging with nitrogen gas prior to analysis. Depending on the pH value of electrolyte, dissolved oxygen is reduced in two steps at the half-wave potentials of about 0 V and –0.9 V, respectively (versus the saturated calomel electrode, SCE) (COLOMBO & VAN DEN BERG 1998).

Step 1

$O_2 + 2\,H^+ + 2\,e^- \rightarrow H_2O_2$ (acidic medium)

$O_2 + 2\,H_2O + 2\,e^- \rightarrow H_2O_2 + 2\,OH^-$ (neutral or basic medium)

Step 2

$H_2O_2 + 2\,H^+ + 2\,e^- \rightarrow 2\,H_2O$ (acidic medium)

$H_2O_2 + 2\,e^- \rightarrow 2\,OH^-$ (neutral or basic medium)

Reduction of oxygen has mainly two effects on the ASV measurement: (i) a large background current, which significantly disturbs the analyte signal, (ii) the change of pH in the vicinity of electrode in a neutral or basic medium, which leads to the precipitation of heavy metal hydroxides (BARANSKI 1987). The latter effect is easy to overcome by using a buffer solution as supporting electrolyte. As for the large background current, although the deoxygenation by purging N_2 is an effective method in laboratory, it is not suitable for in situ measurement. It was reported that interferences due to dissolved oxygen can be ignored or much minimized by using SWASV (WOJCIECHOWSKI & BALCERZAK 1990, BRETT ET AL. 1994, ALFASSI 1994), because at high square wave frequency, the stripping step is less sensitive to irreversible processes, such as the reduction of O_2 (Tercier & Buffle 1993). Incidentally, compared to SWASV, more severe interferences at the presence of dissolved oxygen is found with the differential pulse anodic stripping voltammetry (DPASV), (Wojciechowski & Balcerzak 1990). Thanks to the attractive features of microelectrodes (e.g., high mass transport rate), the influence of dissolved oxygen in the determination of trace metals in water is much less extensive and can be ignored (WANG ET AL. 1993, KOUNAVES & DENG 1993, WANG & TIAN 1992). The coupling of SWASV with microelectrodes is particularly advantageous for the measurement of trace concentrations of heavy metals in the presence of dissolved oxygen.

Figure 40a shows the results of ten repetitive measurements of 1.00 µg/l Cd and Pb in 0.02 M acetate buffer solutions, five with deoxygenation and five without deoxygenation. Except for a current enhancement observed after deoxygenation, no significant differences are observed between the measurements with and without deoxygenation. Figure 40b illustrates another example of simultaneously determination of 4.00 µg/l Zn, 1.00 µg/l Cu, and 0.40 µg/l Cd and Pb in 0.02 M acetate buffer solutions. After deoxygenation the peak baselines of Cd and Pb get smoother, while those of Zn and Cu remain unchanged. This may be due to the fact that at very low concentrations (e.g., 0.40 µg/l Cd and Pb) the influence of dissolved oxygen still has more or less influence on the SWASV measurement. However, at higher concentrations levels (e.g., 4.00 and 1.00 µg/l Zn and Cu), the influence of dissolved oxygen can be ignored.

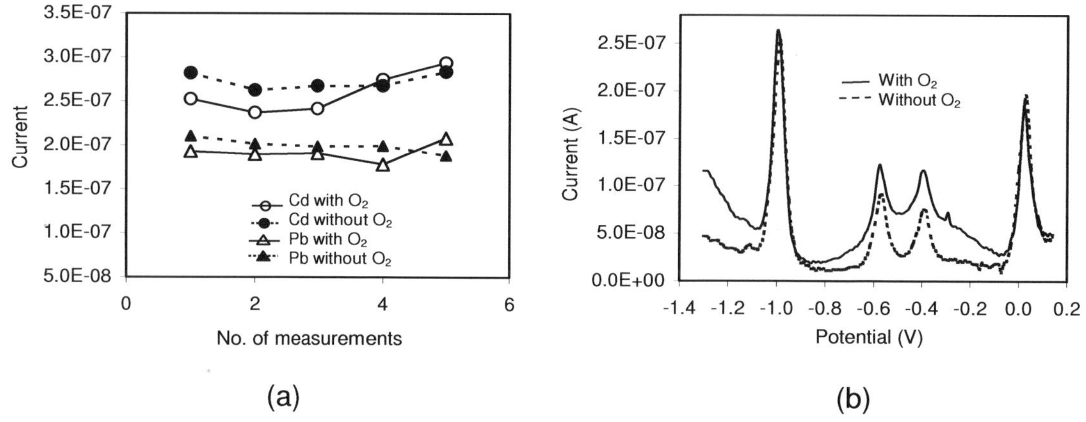

Fig. 40: Influence of dissolved oxygen on the SWASV measurement of heavy metal concentrations in 0.02 M acetate buffer solution (pH = 4.5). Effect on (a) peak current intensity of 1.00 µg/l Cd and Pb, (b) voltammograms of 4.00 µg/l Zn, 1.00 µg/l Cu, and 0.40 µg/l Cd and Pb. Analytical conditions: deposition potential, -1.2 V; deposition time, 180 s; square wave frequency, 300 Hz; amplitude, 25 mV; step potential 15 mV

7.3.2 Dissolved organic carbon (DOC)

It is an understood fact that organic matter, such as humic-, fulvic-, and amino acids, polysaccharides in natural water interferes with the ASV measurement of heavy metals (GOLIMOWSKI & GOLIMOWSKA 1996). Organic matter acts mostly as complexing agent and surface-active substance (surfactant). Organic complexation reduces the reactivity of heavy metals and increases the irreversibility of ASV process, while surface-active organic substances inhibit the access of heavy metals to the electrodes by adsorbing on the mercury surface of electrodes (BREZONIK ET AL. 1976, SAGBERG & LUND 1982, WANG & LUO 1984, BATLEY 1986, VAN DEN BERG 1989). As a result, the presence of organic matter in samples leads in most cases to the depression of current intensity, broadening of the peak, and to shift of the peak potential (WANG 1985). Therefore, prior to the measurement, organic matter has to be destructed in order to free the organically complexed metals and to eliminate the interference of organic surfactants. Instead of wet digestion (GORSUCH 1970), UV-irradiation is commonly used for the breakdown of dissolved organic matter in the determination of trace concentration of heavy metals in natural water, as the chemical contamination is minimized (ACHTERBERG & VAN DEN BERG 1994, ACHTERBERG ET AL. 2001).

In this study, the dissolved organic matter in the river water samples was digested using a self-made UV-decomposition apparatus, which is composed of a high-pressure mercury lamp (UVHQ 125, 125 W, Speciallampen GmbH, Germany), a silica sample loop (the diameter, 1.0 mm; the length, 0.4 m), and a peristaltic pump. After adding of 50 µl of 30% H_2O_2, samples (10 ml for each) were digested in flow mode with a pumping rate of approx 0.4 ml/min and a digestion time of 120 min. For comparison, a commercially available UV-digestion device (the 705 UV Digestor, Metrohm, Switzerland) with a high-pressure mercury lamp (500 W, Efoquarz, Giskon, Switzerland) was used. With this device ten water samples (10 ml for each) were digested simultaneously after the addition of 100 µl of 30% H_2O_2 (Suprapur, Merck) to a 10 ml of sample, using a temperature of 80 °C and a time of 120 min (KOLB ET AL. 1991).

As an example, figure 41a shows the result obtained with a sample digested using self-built UV device. A large background current close to the Zn peak is observed, which is probably caused by the residue of H_2O_2 in solution. Except for the enhanced Zn peak, no substantial improvement was observed by using the self-made UV-digestion device, possibly due to the inefficient digestion of the organic matter. There are a lot of parameters which may influence the efficiency of the UV-digestion, including the type and power of the mercury lamp, digestion tube (material), oxidant, digestion time, temperature, etc.. Mercury vapour lamps of various constitutions (low-, medium-, and high pressure) with different line spectra (200 ~ 435 nm) were utilized as UV sources (KOLB ET AL. 1992, PISCH ET AL. 1993, YOKOI ET AL. 1999). Low-pressure mercury lamps have been reported to be more effective in the decomposition of organic compounds, especially for aromatic compounds, due to their much higher light intensity at short wavelengths (185 nm and 254 nm) (YOKOI ET AL. 1995, YOKOI ET AL. 2002). The spectrum line at 185 nm is particularly important in producing hydroxyl radicals, and thus in decomposing the organic matter (YOKOI ET AL. 1999). The poor results obtained with the self-made UV digestion device may be due chiefly to the inadequacy of the mercury lamp used. However, other parameters like the transparency of the silica loop to UV-light, the distance between lamp and silica loop, etc. should be considered, too.

As shown in figure 41b, the results obtained with the digestor from Metrohm were also not satisfactory. Though the peak current of Pb is much improved and even a slight Cd peak generates after UV decomposition, apparently, the peak currents and peak shapes of Zn and Cu got much worse. It seems that UV-digestion is more efficient for Cd and Pb. It is difficult to give the exact reasons for the ineffective UV-digestion even with the commercial UV-digestor. However, it is undoubtedly that H_2O_2 has a large influence on the efficiency of the UV-digestion. The addition of H_2O_2 increases the digestion efficiency by producing more hydroxyl radicals ($HO^\bullet$), as the concentration of hydroxyl radicals in solution generated from water may be not sufficient (SAUR 1992, PISCH ET AL. 1993).

$H_2O + h \cdot \nu \rightarrow H^\bullet + HO^\bullet$

$H_2O_2 + h \cdot \nu \rightarrow 2\ HO^\bullet$

It was suggested that in order to aid the breakdown of organic matter, a concentration of 10 mM H_2O_2 should be added to the sample prior to UV-digestion (ACHTERBERG & BRAUNGARDT 1999). On the other hand, the addition of too much H_2O_2 may exert a negative influence on the SWASV measurement of trace metal concentrations (YOKOI ET AL. 1995, YOKOI ET AL. 1999), especially for water samples with low DOC contents. Therefore, the amount of H_2O_2 added should be matched with the DOC content of the sample. If the DOC content in sample is very low, too much addition of H_2O_2 should be avoided, or if necessary, the addition of H_2O_2 is even simply avoided (SAUR ET AL. 1980).

Literature data indicate that by using medium exchange and standard addition, the interference of organic matter on the SWASV analysis of trace metal concentrations can be minimized (WANG 1985, WAHDAT & NEEB 1987, DANIELE ET AL. 1989, SILVA ET AL. 2001). In the acidified samples, the organic complexation of metals is reduced to a large extent (VAN DEN BERG 1989). After filtration and acidification the river water samples used in this study were directly analysed without UV-digestion, but with electrolyte exchange. Satisfactory results were obtained, especially for Cd and Pb. Results indicate that if the DOC content is very low it is possible to determine trace concentrations of heavy metals in natural waters without destruction of the dissolved organic matter. Actually, for the in situ and on site measurement of metal species, neither UV-digestion nor acidification of samples is allowed.

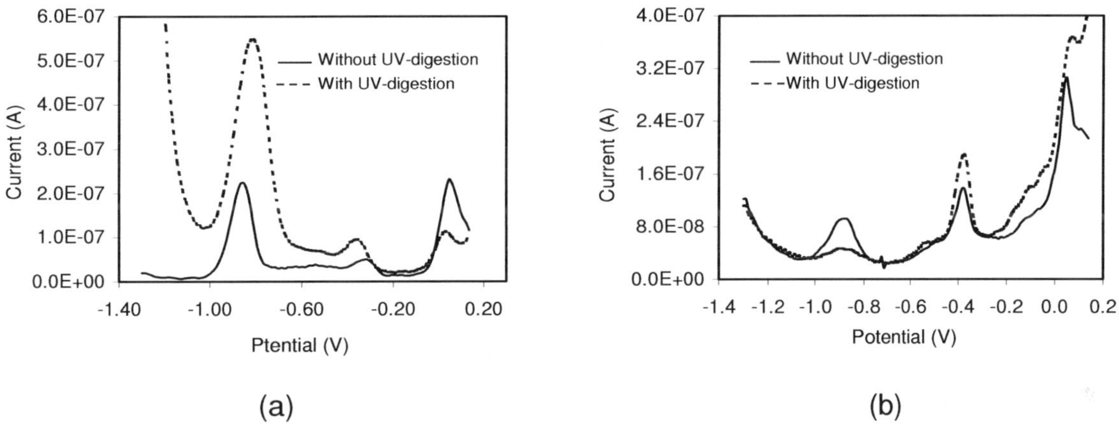

Fig. 41: Comparison of SWASV measurement of heavy metal concentrations in river water samples with and without UV-digestion: (a) self-made UV apparatus, (b) 705 UV-digestor of Metrohm. Analytical conditions: deposition potential, -1.4 V; deposition time, 240 s; square wave frequency, 300 Hz; amplitude, 25 mV; step potential 15 mV

7.3.3 Intermetallic compounds

It is reported that Cu-Zn intermetallic compounds (e.g., CuZn, $CuZn_2$, $CuZn_3$) may interfere with the simultaneously determination of the concentrations of Cu and Zn in aqueous solutions (SHUMAN & WOODWARD 1976, PICCARDI & UDISTI 1987). During the stripping step of the ASV, the share of Zn entering the Cu-Zn intermetallic compounds strips at a potential close to that of Cu, decreasing the peak current of Zn and increasing the peak current of Cu. Several approaches have been suggested to eliminate the formation of Cu-Zn compounds, such as standard addition (COPELAND ET AL. 1974, GERLACH & KOWALSKI 1982), the addition of a complexing agent (MARQUES & CHIERICE 1991), and the addition of a third element (WANG ET AL. 1984, BRETT ET AL. 1996), etc.. Among those the addition of a third element like Gallium was recommended (NEIMAN ET AL. 1980, WANG ET AL. 1999). Compared to Zn Ga forms a stronger intermetallic compound with Cu, thus the formation of the Cu-Zn compounds is suppressed, making it possible to determine the concentrations of Zn in the presence of an excess of copper ions. In this study, because the low concentrations of Cu and Zn in the samples (Zn 4 ~20 µg/l, Cu 0.5 ~ 2.5 µg/l), no obvious interference of Cu-Zn compounds was observed in the SWASV measurement by using standard addition. This is in agreement with literature data that the Cu-Zn interference is less of a problem for samples with low concentrations of Cu and Zn in natural waters (BELMONT-HEBERT ET AL. 1998, HERDAN ET AL. 1998, PEI ET AL. 2000).

7.3.4 Chloride ions

Chloride ions may influence the ASV measurements of heavy metals due to the formation of insoluble calomel (Hg_2Cl_2) on the mercury film electrodes (WU 1994, JAGNER ET AL. 1996, NOLAN & KOUNAVES 2000).

$Hg + Hg^{2+} + 2Cl^- \rightarrow Hg_2Cl_2$

The effects were also reported with microelectrodes. Because of the accumulation of Hg_2Cl_2, the surfaces of microelectrodes become degraded in Cl^--rich media and the lifetime of microelectrodes got shortened (WANG & TIAN 1992, NOLAN & KOUNAVES

(a) (b)

Fig. 42: Comparison of the SWASV measurements of 1.00 µg/l Cd and Pb in Cl^--free solutions and in 3% NCl solutions. Influence on (a) peak current intensities of four consecutive measurements, (b) voltammograms. Analytical conditions: deposition potential, -1.2 V; square wave frequency, 300 Hz; amplitude, 25 mV; step potential 15 mV

1998, NOALN & KOUNAVES 1999). In order to investigate the influence of chloride ions on the Ir-UMEA chip, the SWASV measurements were carried out in synthetic NaCl solution (3 wt.%) and in natural seawater, with both of the solutions buffered to 0.02 M acetate. Figure 42a and 42b show the peak current signals generated in 1.00 µg/l Cd and Pb of Cl^-- free 0.02 M acetate buffer solutions and in synthetic NaCl solution (3 wt.%). The higer current intensities measured in the NaCl solution may be due to higher blank values of chemicals, while the shift of peak potentials to more negative values is possibly caused by the change of pH or by the complexation of metal ions in the presence of chloride ions. Peak potential is usually subject to a shift towards more negative values if the metals are complexed (SKOOG & LEARY 1996). According to the Nernst equation, the half wave potential depends on the concentration of free ions in solution. After complexation, the amount of free ions in solution is reduced, and consequently, the peak potential is more negative.

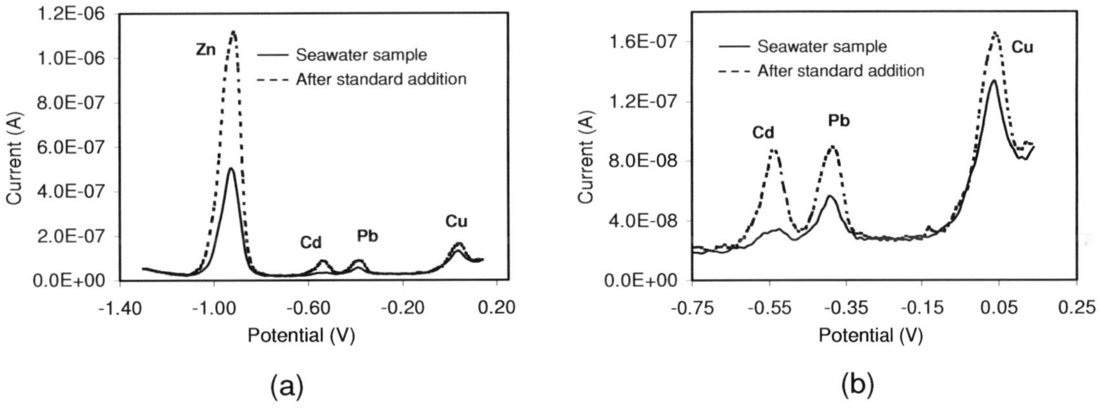

Fig. 43: Example of simultaneously SWASV determination of Cu, Pb, Cd, and Zn in seawater by successive addition of 10 µg/l Zn, 2.5 µg/l Cu, and 0.4 µg/l Cd and Pb in 0.02 M acetate buffer solutions. Analytical conditions: deposition potential, -1.4 V; deposition time, 240s; square wave frequency, 300 Hz; amplitude, 25 mV; step potential 15 mV

Seawater samples were successfully measured by means of UMEA sensors using SWASV method. The voltammograms of measurements are shown in figure 43a and 43b, indicating well-defined peaks, good peak resolution, smooth baseline, and high sensitivity. No apparent deterioration of UME surfaces was observed in the Cl^--rich media (synthetic and natural), proving that the UMEA sensor may be successfully used for the determination of trace metals in seawater. In fact, in this study more satisfactory measurement results were obtained in seawater than in river water. This is in agreement with the literature data, which document the successful application of both conventional electrodes and microelectrodes in the trace metal analysis of seawater (GRASSHOFF ET AL. 1999, TERCIER-WAEBER ET AL. 1999, DANIELE ET AL. 1989).

8. Application of ultramicroelectrode array in natural waters

8.1 Materials and methods

8.1.1 Sampling

The area of investigation is located in the Baden-Wuerttemberg, in the South-West of Germany. The sampling stations are distributed over the catchment area of the river Rhine including tributary streams and rivers (Tab. 4). The samples were filtered on site through 0.45 µm acetate membrane filter, preserved at pH < 2 with concentrated HNO3 (Suprapur), and stored in refrigerator at approx. 4 °C before laboratory analyses.

8.1.2. Apparatuses and reagents

The concentrations of heavy metals in river water samples were determined by means of the Ir-UMEA sensor and high resolution inductively coupled plasma mass spectrometry (HR-ICP-MS) (VG Axiom, VG Elemental, England), respectively. The content of dissolved organic carbon (DOC) was measured by means of the DIMA-TOC 100 Analyser (DIMATEC, Essen, Germany) with potassium hydrogen phthalate (analytical grade, Merck) as DOC standard. Field parameters like pH, dissolved oxygen, conductivity, and temperature were measured with a multi-meter (Multi 350i, WTW, Germany). For the HR-ICP-MS measurements, calibration solutions were prepared from the ICP-MS multi-element standard (CPI, Amsterdam).

Tab. 4: Sampling stations, related rivers, and sample numbers

Number	Station	River
P 1	Russheim	Pfinz
P 2	Russheim	Saalbachkanal
P 3	Philippsburg	Saalbach
P 4	Mündung	Wagbach
P 5	Altussheim	Krigbach
P 6	Mannheim	Rhine
P 7	Mannheim	Neckar
P 8	Riegel	Dreisam
P 9	Riegel	Elz
P 10	Vogelgrün	Rhine
P 11	Breisach	Rhine
P 12	Jamborschwelle	Leopeldskannal
P 13	Rust	Elz Rust
P 14	Nonnenweier	Schuttterkanal
P 15	Steinmauern	Murg
P 16	Iffezheim	Sandbach
P 17	Helmlingen	Rench
P 18	Kehl	Kinzig
P 19	Offenburg	Kinzig
P 20	Oberkirch	Rench

8.1.3 Analytical procedures

Prior to measurement, the concentration levels of heavy metals in water samples were estimated and determinations in synthetic solutions were performed, in order to establish suitable SWASV parameters (e.g., deposition time, deposition potential, etc.) and to determine the appropriate metal concentrations for the standard addition. The samples were not treated by UV-digestion but were buffered to 0.02 M acetate solutions (pH = 4.5). After the preconcentation step at a given time (120 ~ 360 s) and a controlled deposition potential (-1.4 V), electrolyte was exchanged to a pure 0.02 M acetate buffer solution and the electrodes were washed for about 30 s by maintaining the same deposition potential. Then, the stripping process was carried out without equilibration. Between two measurements, a cleaning step was intercalated using amperometry at a potential of - 0.05 V for 60 s, in order to avoid memory effects. In most cases, concentrations of Cu, Pb, Cd, and Zn were simultaneously determined by standard addition, spiking the samples with 4.00, 8.00 µg/l of Zn, 1.00, 2.00 µg/l of Cu, and 0.20, 0.40 µg/l of Cd and Pb, respectively. Due to the high concentrations of Zn and Cu, and the very low concentrations of Pb and Cd, the four elements were determined separately in some of the samples. Standard addition was preferred as the sensitivity of the stripping voltammetric analysis may vary between samples with different ionic strength and containing different concentrations of dissolved organic matter (ACHTERBERG & BRAUNGARDT 1999). The quantitative evaluation of the trace metal concentrations in samples is based on the linear correlation between peak current intensities and concentrations. According to this relationship the concentrations can be evaluated by the following formula (DONG & KE 2000):

$$C_x = \frac{C_s V_s h_x}{H(V_x + V_s) - h_x V_x}$$

where C_x, the concentration of heavy metal in sample (µg/l); C_s, the concentration of heavy metal in the standard solution (µg/l); V_x, the volume of sample (ml); V_s, the added volume of standard solution (ml); h_x, the peak current intensity of sample; H, the peak current intensity after a successive standard addition.

For comparison and validation, the river water samples were also analysed by means of HR-ICP-MS, and the total concentrations of trace metals were obtained. The instrumental and acquisition parameters are listed in Table 5.

Tab. 5: The measurement parameters of HR-ICP-MS

Instrument	VG Elemental (Axiom)
Forward power	1350 W
Reflected power	< 20 W
Sampling uptake rate	1 ml/min
Scan mode	Magnet scan
Torch	Quartz torch (Glass Expansion, Australia)
Nebulizer	Type Meinhardt (Glass Expansion, Australia)
Spray chamber	Conical impact wall (bead)
Temperature of spray chamber	10 ± 1 °C
Argon gas flow rates (l/min)	Cool gas: 14.2
	Auxiliary gas: 0.65
	Nebulizer gas: 0.94
Data acquisition parameters	
Acquisition mode	Pulse counting
Number of runs	3
Dwell time	25 ms
Points per peak width	20
Sweep	1
Internal standard	^{103}Rh

8.2 Results and discussion

8.2.1 Preliminary tests

Figure 44 shows the results of simultaneously determination of trace meals in 0.02 M acetate buffer solutions using the Ir-UMEA sensor, with successive addition of 2.00, 4.00, 6.00 µg/l of Zn, 0.50, 1.00, 1.50 µg/l of Cu, and 0.20, 0.40, 0.60 µg/l of Cd and Pb. Excellent voltammograms are obtained with proportional and well-defined peaks, and smooth baselines. Shown are also the peak potentials: Zn, -0.99 V, Cd, -0.58 V, Pb, -0.39 V, and Cu, +0.029 V, which are typical values in a non-complexed medium (BRENDEL & HOFFSTETTER-KUHN 1996). It can be concluded that (i) the selected electrochemical parameters, such as deposition time, deposition potential, are suitable, (ii) the supporting electrolyte is adequate for the simultaneously determination of Cu, Pb, Cd and Zn, (iii) at the considered concentration levels, intermetallic interferences (e.g., Cu-Zn compounds) are negligible, (iv) at the deposition potential of -1.4 V a slight hydrogen peak close to the Zn peak occurs, which is however less problematic for the measurement. Therefore, it is concluded that the established measurement conditions can be applied in natural water samples.

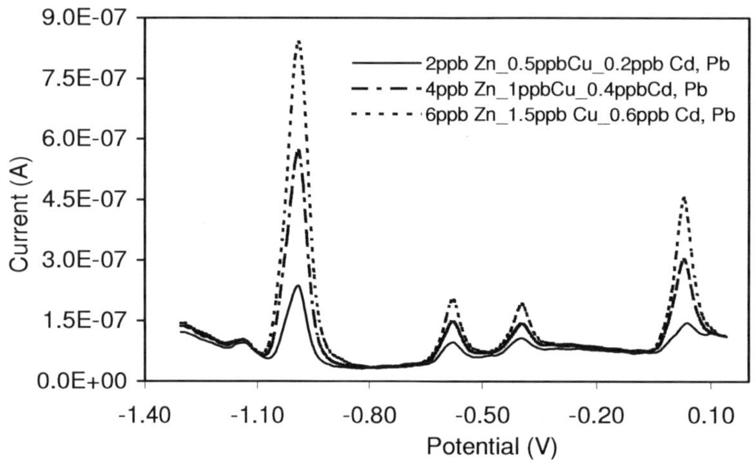

Fig. 44: The SWASV voltammograms of Cu, Pb, Cd and Zn obtained in synthetic aqueous solutions. Measurement conditions: deposition potential, –1.4 V; deposition time, 240 s; frequency, 300 Hz; amplitude, 25 mV; step potential, 15 mV; electrolyte, 0.02 M acetate buffer (pH = 4.5)

8.2.2 Measurement of the river water

As examples the voltammograms obtained with the samples at stations Helmlingen (p17) and Kehl (p18) are shown in figure 45 and figure 46. The well-defined peaks of Zn, Cu, Pb, and Cd for both of the samples and the proportional increment in peak currents for sample p17 by successive addition of 4.00, 8.00 µg/l Zn, 1.00, 2.00 µg/l Cu, and 0.2, 0.4 µg/l Cd and Pb (Fig. 45) provide the evidence that the SWASV measurements were not significantly affected by organic or complexing substances in acidified samples. However, compared to the measurements in synthetic solutions, a shift of peak potentials towards more positive values (Zn, -0.88 V, Cd, -0.55 V, Pb, -0.38 V, and Cu, +0.039 V) was observed for all the four metal ions (especially for Zn with a shift of 0.11 V). Slight peak broadening was found with Zn, which is probably

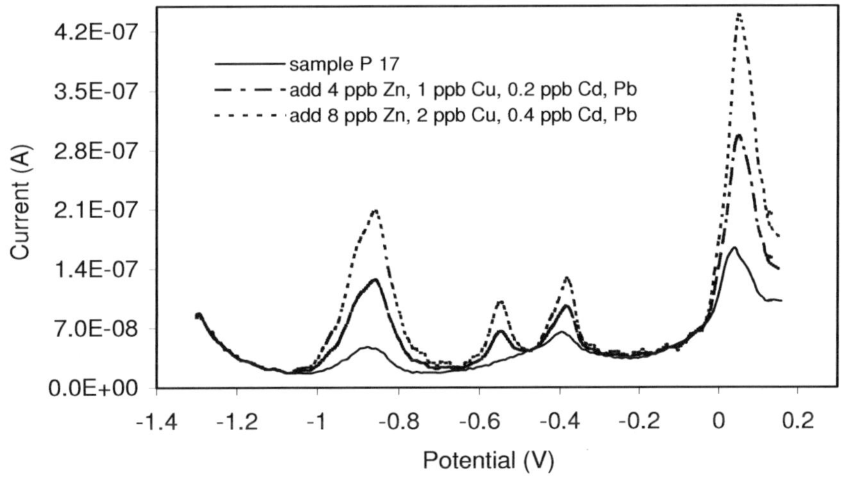

Fig. 45: The SWASV voltammograms of Cu, Pb, Cd, and Zn obtained in the river water sample of the station Helmlingen (P17). Measurement conditions: deposition potential, -1.4 V; deposition time, 240 s; frequency, 300 Hz; amplitude, 25 mV; step potential, 15 mV; electrolyte, 0.02 M acetate buffer (pH = 4.5)

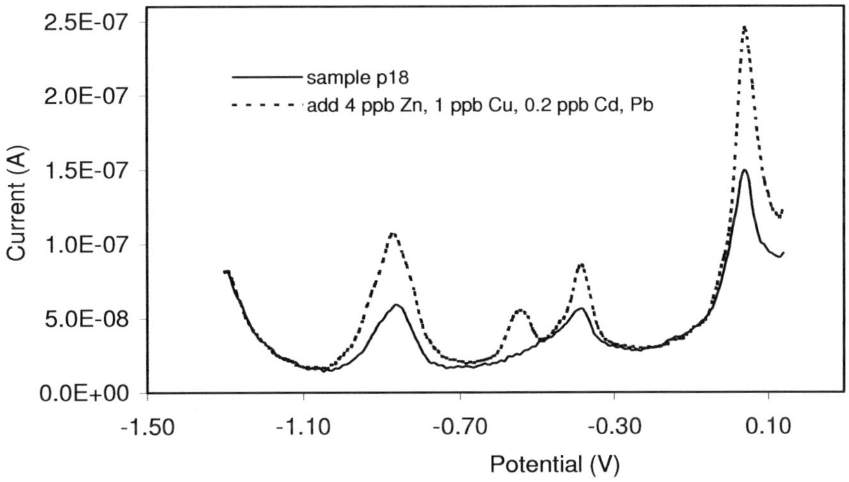

Fig. 46: The SWASV voltammograms of Cu, Pb, Cd and Zn obtained in the river water sample of station Kehl (p18). Measurement conditions are identical to the figure 45

caused by the increased irreversibility of the stripping process at the presence of surface-active substances in the samples (WANG 1985). This indicates that although the influence of organic matter on the SWASV measurement can be greatly minimized by means of medium exchange and standard addition (VAN DER BERG 1989), it can still not be ignored, especially for very reproducible measurements of heavy metals in natural waters.

8.2.3 Comparison with HR-ICP-MS

The concentrations of Cu, Pb, Cd, and Zn measured in the river samples by means of the Ir-UMEA sensor using SWASV and HR-ICP-MS are listed in appendix 2. In general, compared to HR-ICP-MS, the measurements with the Ir-UMEA sensor yielded lower concentrations. This difference can be explained at least in part by the total different analytical approaches. The HR-ICP-MS analysis provides total metal concentrations including also metals in complexed form, while the ASV approach measures in principle merely the concentrations of mobile species (free ions and small labile complexes), provided the sample is not acidified and the organic matter in sample is not destructed (BUFFLE ET AL. 1997). It was reported that the dissolved fraction of trace metals obtained by in-situ ASV measurement accounts for only 10-30% of the total metal concentrations measured by means of ICP-MS, or by ASV with UV-digestion (TERCIER-WAEBER ET AL. 1999). Even though several measures were taken in this study to minimise the influence of organic matter, the concentrations of trace metals measured by ASV without UV-digestion are in most cases still lower than those obtained with ICP-MS.

The relative difference between the concentrations measured by HR-ICP-MS and UMEAs sensor was expressed as:

$$D\ [\%] = \frac{C_{ICP} - C_{UMEA}}{C_{ICP}}\ \text{x}\ 100\%$$

D [%], relative difference; C_{ICP}, the concentration measured by HR-ICP-MS; C_{UMEA}, the concentration measured by the Ir-UMEA sensor.

The best agreement of the two methods was found with Pb. Figure 47a shows the correlation between the concentrations of Pb obtained with the UMEA sensor and with HR-ICP-MS. For 60% of the stations, the relative differences lie in the range of +21.3% ~ -22.8%. For 30% of the stations the relative difference is between +42.5 % and -32.8 %. In most samples (80%), the concentrations of Pb are less than 0.10 µg/l, and there are only four samples with the heavy metal contents larger than 0.2 µg/l. For such low concentration levels, relative differences of less than 40% are quite acceptable.

The concentrations of Cd in the river water samples are extremely low and it is difficult to be accurately determined using both of the methods. The results obtained

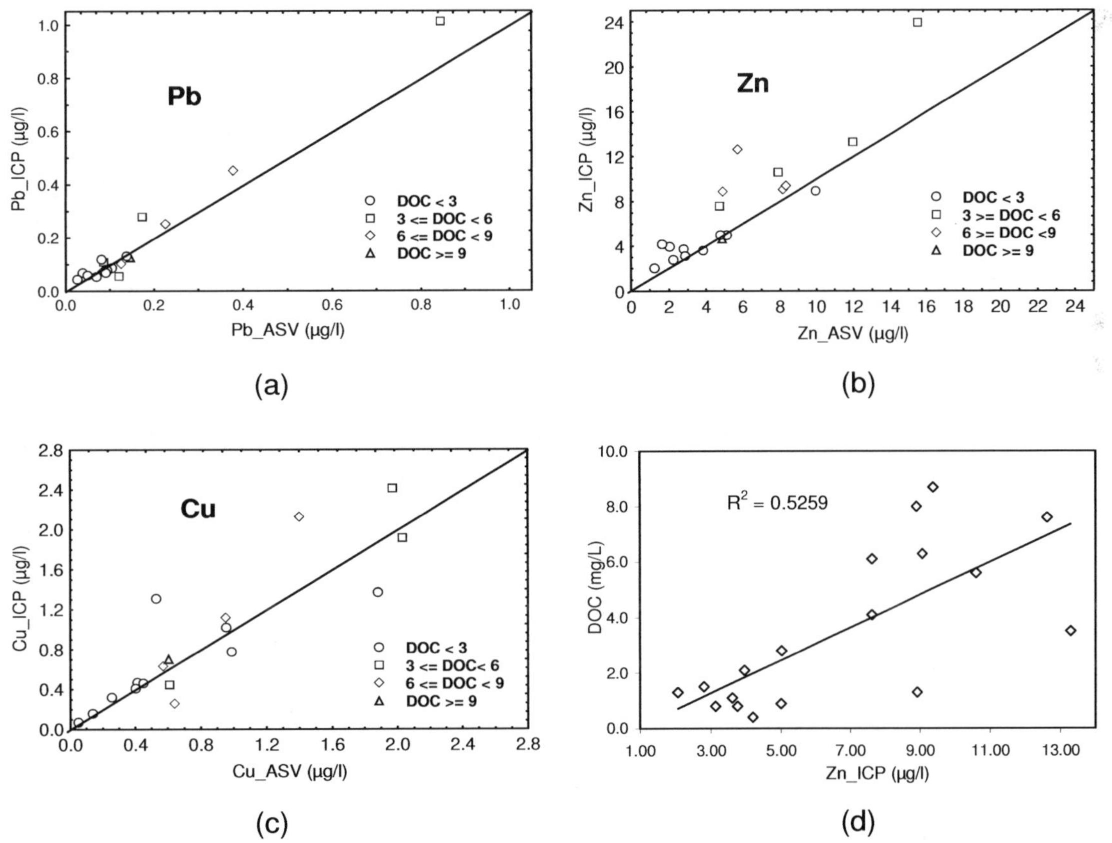

Fig. 47: Correlation between the concentrations of heavy metals measured with the UMEA sensor and HR-ICP-MS, (a) Pb, (b) Zn, (c) Cu, (d) correlation between DOC contents and Zn concentrations obtained with HR-ICP-MS

by means of HR-ICP-MS show that for approx 75% of the samples the concentrations of Cd are lower than 0.025 µg/l, which is below or near the detection limit of this element.

The concentrations of Zn determined with the UMEA sensor are generally lower than the results obtained with HR-ICP-MS. As shown in figure 47b, for only 45% of the samples, the concentrations of Zn by the UMEA sensor are in good agreement with those by HR-ICP-MS, with the relative differences of ± 11.6%. About 30% of the samples show the relative differences between +20.4% and +39.2%, indicating a relative large deviation of the two methods. The relative large differences are found with four samples (+44.8 ~ +60.7%), with apparently higher concentrations of HR-ICP-MS.

Concerning Cu, good correlations were found in more than half of the samples (55%), with relative differences between 1.4% and 20.0%. Relatively low concentrations were found with five samples, with relative differences in the range of +34.2% and –37.5%. Extremely high differences were found for four samples, in which either the ICP-MS results were significantly higher or the measurements were failed carried out using the UMEA sensor. In order to find the causes of the inconsistencies of the two methods, apart from trace metals, parameters like dissolved organic carbon (DOC), pH, dissolved oxygen, conductivity, and temperature were additionally measured (appendix 3). As an example the effect of DOC contents on the measurements of heavy metals is given. An obvious positive correlation between the concentrations of Zn measured by HR-ICP-MS and DOC contents is shown in figure 47d, with an correlation coefficient of r = 0.73. The correlation suggests physicochemical interaction between Zn and organic matter in river water, such as complexation or adsorption, which would inhibit the mobility of Zn ions. As a result, the measurements of Zn using the UMEA sensor are disturbed by interferences, which are supported by the evidently low concentrations. In contrast, no obvious correlation was found between the DOC contents and the Pb and Cu concentrations (Fig. 47a, 47c). The large differences observed with Cu in several samples may be due to (i) the ability of Cu to bind strongly with many organic and inorganic ligands; (ii) the existence of Cu in more than one oxidation state; (iii) the occurrence of the stripping peak of Cu at a potential where many organic species are adsorbed on the mercury electrodes; (iv) the low solubility of Cu in mercury; (v) intermetallic compounds (e.g., Cu-Zn compound), and (vi) the irreversibility of the Cu (II) reduction (BATLEY 1986, HERDAN 1998). The satisfactory results for Pb and Cd obtained with the UMEA sensor suggest that the presence of organic matter is less problematic in the SWASV measurements. In fact, compared to Zn and Cu, Pb and Cd possess a more stable electrochemical behavior in the ASV measurement. For both of the elements, influences resulted from both hydrogen evolution at more negative deposition potentials and from the oxidation of mercury at more positive deposition potentials can be excluded. As for parameters, for instances, pH, dissolved oxygen, conductivity, and temperature, no correlation was observed with heavy metals.

8.2.4 Regional distribution of the concentrations of heavy metals

The regional distribution of Cu, Pb, and Zn concentrations in river water is shown in figures 48, 49, and 50. Besides the distribution of metal concentrations, the figures allow at the same time a comparison of the trace metal concentrations obtained with the UMEA sensor and HR-ICP-MS. The size of circles is proportional to the sum of the square roots of the results obtained with the two methods. Equal segments (red, ICP-MS; green, UMEA sensor) mean that the results of the two methods are in good agreement.

For Pb, an anomalously high concentration was found at the station Russheim in river Pfinz (p 1) (1.01 µg/l, ICP-MS and 0.85 µg/l, the UMEA sensor) (Fig. 48). Relatively high concentrations were registered at Philippsburg (p3), Russheim in river Saalbachkanal (p2), and Steinmauern (p15), with an average concentration of 0.32 µg/l (ICP-MS), and 0.26 µg/l (the UMEA sensor), respectively. The slightly elevated Pb concentrations are found in urban areas, such as Mannheim, Karlsruhe, and Heidelberg, which may be explained by the increased anthropogenic Pb input. However, at most of the stations (80%), the Pb concentrations are very low (< 0.10 µg/l).

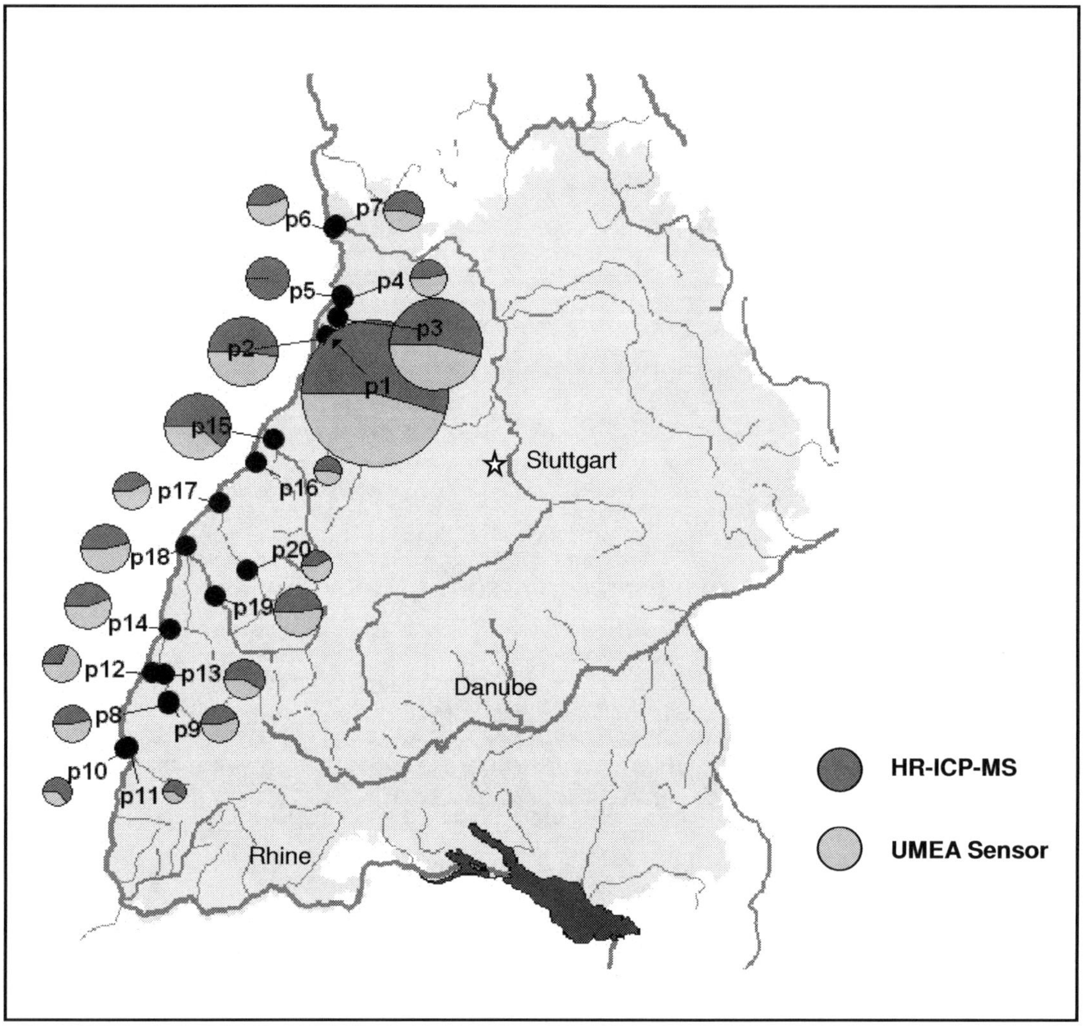

Fig. 48: Regional distribution of the Pb concentrations in the investigation area

The regional distribution of Zn concentrations is shown in the figure 49. The highest concentration was found at the station Jamborschwelle (p12) (23.93 µg/l, ICP-MS; 15.52 µg/l, the UMEA sensor). Increased Zn concentrations were found at the stations Russheim in the river Pfinz (p1), Nonnenweier (p14), Mannheim (p7), Mündung (p4), Philippsburg (p3), and Riegel in the river Dreisam (p8) with the average concentrations of 10.66 µg/l (ICP-MS) and 8.67 µg/l (the UMEA sensor), respectively. At most of the stations (60%) the Zn concentrations are very low with the average concentrations of 4.56 µg/l (ICP-MS) and 3.43 µg/l (the UMEA sensor), correspondingly. There are two areas with slightly increased Zn values: one overlaps with the area of the higher Pb contents around Karlsruhe, Mannheim, and Heidelberg; the other one is near Freiburg.

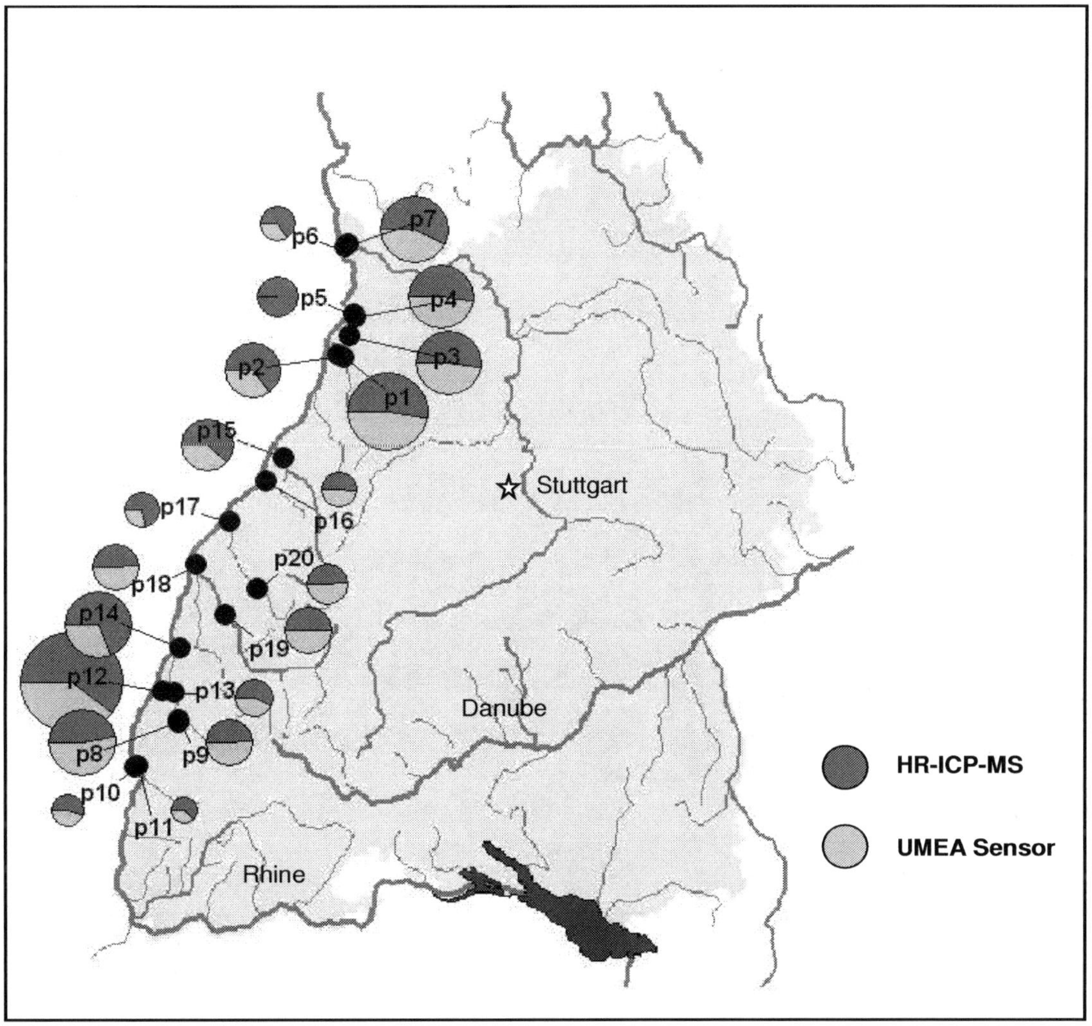

Fig. 49: Regional distribution of the Zn concentrations in the investigation area

The regional distribution of the Cu concentrations is illustrated in figure 50. The highest concentrations are located at the stations Mannheim (p7), Nonnenweier (p14), Jamborschwelle (p12), and Offenburg (p19) with the average concentrations of 1.96 µg/l (ICP-MS) and 1.82 µg/l (the UMEA sensor), respectively. Slightly lower concentrations occur at Mannheim (p6), Philippsburg (p3), Helmlingen (p17), and Oberkirch (p20), with the average Cu concentrations of 1.06 µg/l (ICP-MS) and 0.86

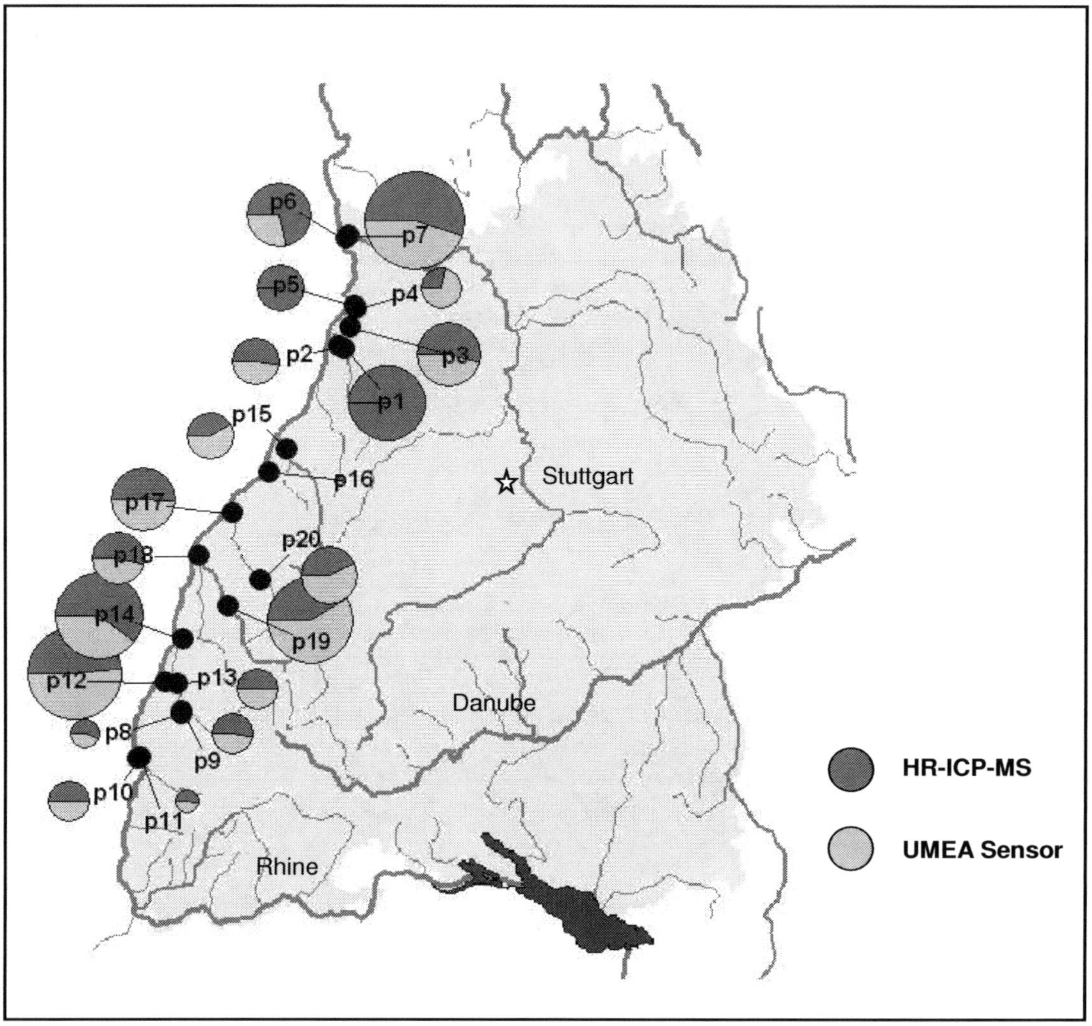

Fig. 50: Regional distribution of the Cu concentrations in the investigation area

µg/l (the UMEA sensor), respectively. The distribution pattern of Cu is similar to that of Zn. It should be noted that the highest Cu content of 2.28 µg/l measured with ICP-MS at Russheim in the river Pfinz (p1) was not measurable with the UMEA sensor. Similarly, at station Altussheim, none of the four elements could be measured with the UMEA sensor, which may be due to higher contents of surfactants in the samples.

8.3 Conclusion

In general, the determination of heavy metal concentrations in river water samples using the UMEA sensor yielded relative satisfactory results. Though the organic matter in samples was not destructed and no deoxygenation was performed, in more than half of the samples the heavy metal concentrations measured by the UMEA sensor are in good agreement with those obtained with HR-ICP-MS, with relative differences less than 20%. This difference is even less significant, if we take into consideration that the measured concentrations are very low. Compared to Pb and Cd, a more severe interference of organic matter was observed with Zn and Cu, as confirmed by the shift of the peak potentials in the voltammograms and by the larger deviation in the metal concentrations measured using the two methods. However, by means of medium exchange and standard addition, the interferences caused by organic matter can be reduced to a large extent. Higher concentrations of heavy metals in river water were found at the areas near large cities and with more industries.

9. Summary and prospects

A novel electrochemical sensor has been developed for the determination of trace concentrations of heavy metals in water. The sensor is based on iridium ultramicroelectrode arrays (UMEAs), which have been fabricated using the modern microphotolithography of thin-film technique. The UMEA chip with a size of 6.4 x 5.0 mm is made up of 4048 individual microelectrodes as working electrodes, arranged in four separate arrays. Each array consists of 46 x 22 = 1012 disc shaped microelectrodes with a diameter of 1.8 µm each, an interelectrode distance of 25 µm, and a recess depth of 0.2 µm. The total electrode area of each array is 2575 µm^2. Electrochemical characteristics, analytical performance, influence parameters, and the application potential of the newly developed Ir-UMEA sensor have been systematically investigated in this study.

Compared to macroelectrodes, microelectrodes possess a lot of attractive properties, such as high mass transport, immunity to ohmic drop, reduced changing current, and small RC constant, enhanced signal-to-noise ratio, etc., which open new possibilities in analytical chemistry. However, the drawback of individual microelectrodes is the exceptionally low measurement current, which is difficult to measure with current instrumentation. This may be overcome by the construction of microelectrode arrays. In this way, the advantageous features of individual microelectrodes can be maintained, if arrays with a proper geometry are designed and high quality chips are fabricated. The quality of UMEA chips has been checked by surface analysis using scanning electron microscope (SEM), environmental scanning electron microscope (ESEM), and atomic force microscope (AFM), respectively. The results show that the iridium electrodes have very regular disc shapes, well-defined areas, sharp edges, and smooth surfaces, and are free from residues of photoresist or passivation layer, indicating a high fabrication quality. Although small cracks can be seen on the surface of UMEs under the large magnification of ESEM (40,000 X), no influence on the measurements has been observed with those electrodes.

The electrochemical behavior of the UMEAs has been studied by means of cyclic voltammetry (CV). Compared to the cyclic voltammetric behavior of individual microelectrodes (i.e., sigmoidal voltammograms at slow scan rates, and peak shaped the cyclic voltammetric behavior of the electrode arrays is more complicated. Depending on the way the microelectrodes are arranged in the arrays (geometry) or on the scan rate of the experiments (timescale), the same array electrodes may exhibit the cyclic voltammetric behavior of micro-, quasi micro-, or even macroelectrodes. There is no consensus on the optimum ratio of the interelectrode distance (w) to the electrode diameter (d) to avoid the shield effect caused by the overlap of the diffusion layers. The cyclic voltammetric behavior of the UMEAs used in this study has been investigated by varying the scan rate from 5 mV/s to 30 V/s. Overall, sigmoidal cyclic voltammograms have been obtained, suggesting that at the given packing density of the UMEAs (i.e. a w/d ratio of 14) the individual diffusion fields of the electrodes remain isolated within the timescales used, and that a steady state with dominantly radial diffusion is attained. However, slight deviations from the microelectrode behavior have been observed at very low (e.g., 5 mV/s) and very high (e.g., 30 mV/s) scan rates, which are supported by the little changes in the shape of the cyclic voltammograms and in the intensity of the measurement currents. This indicates a transition of mass transport from radial diffusion to linear diffusion. In addition to cyclic voltammetry, the electrochemical behavior of the UMEAs has also been examined by using chronoamperometry (CA). In principle, depending on the experimental timescales three different regimes can be distinguished concerning the mass transport of array electrodes. At very short or very long timescales, linear diffusion is dominant and the electrodes are in the Cottrell state, while at intermediate timescales, radial diffusion prevails and the electrodes are in the steady or quasi-steady state. In this study, the amperometric response of the UMEAs has been recorded at a timescale from 0.001 to 1000 s. Generally, only two diffusion states have been observed: the Cottrell state, and the steady or quasi-steady state, which are proved by an obvious current decrease at the very beginning of the experiment and by a roughly constant current intensity for the rest of the time, respectively. It means that during the given experimental time (1000 s) the second Cottrell state which is expected at very long times was not really attained, demonstrating the microelectrode behavior of the UMEAs. Compared to values generated by the digital simulation model of Shoup and Szabo (1984), much higher and more stable amperometric response has been found with the UMEAs, which is probably due to the unsuitability of this empirical model to the array electrodes.

The analytical performance of the UMEAs has been investigated by determining the trace concentration of heavy metals (Cd, Pb, and Cu) using the square wave stripping voltammetry (SWASV) in synthetic solutions. It has been found that the quality of the mercury film plays a decisive role in the measurements. Instead of chronoamperometry (current vs. time), chronocoulometry (charge vs. time) was applied for mercury plating, because the charge is a better criteria to determine the duration of the plating period in getting a mercury film of constant geometry. For four arrays, the optimum charge has been found to be in the range of 0.20 ~ 0.30 mC. Lower plating charge causes a decrease in the measurement sensitivity, while higher plating charge results in instable mercury films. The analytical performance of UMEAs (e.g., sensitivity, precision, and accuracy) has been examined using an optimal mercury film. For a given mercury film, the sensitivity of the measurement has been found to be proportional to the deposition time and the square wave frequency. Concentrations of Cd and Pb down to 0.25 µg/l in 0.02 M acetate buffer solution (pH = 4.5) can be easily determined with a deposition time of about 5 min and a frequency of 300 Hz. The reproducibility of the measurements has been examined by repetitive measurements of 1.00 µg/l Pb and Cd. Based on the 123 measurements of Cd, the relative standard deviation of 11% has been found, while for the 112 measurements of Pb, the relative standard deviation for different film varies from 9% to 13%. The accuracy of measurements has been tested with the reference sample (Trace Metals in Drinking Water, HPS, USA). For 14 repeated measurements, the results were within ± 6% of the certified value of 1.00 µg/l for Cd and within ± 9% of the certified value of 1.00 µg/l for Pb. An Ir-UMEA chip can be used for the generation of a mercury film (plating) for at least 10 times, corresponding to approximately 500 measurements of heavy metals in synthetic aqueous solutions. After that it is difficult to generate mercury films of good quality using the same UMEA chip, which is probably due to damaged electrodes, as supported by the EDX analysis. The roughness analysis of the electrode surfaces by means of atomic force microscope (AFM) shows that the electrodes are getting polluted with time, and consequently, the electrodes stop to work properly. Since the formation of calomel (Hg_2Cl_2) may be excluded during the experiments, the only conceivable source for fouling is fine plastic particles caused by the pumping tube, the sealing O-ring, or the flow through cell, which may however be effectively cleaned by using an ultrasonic water bath. Compared to a single array, the four arrays used together do not increase the measurement sensitivity, but in this way, the lifetime of the UMEAs can be prolonged.

The SWASV measurement using the UMEA sensor is generally influenced by instrumental parameters, experimental conditions, and interferences. The effects of the square wave parameters on the measurements have been investigated in detail. Considering the intensity of the peak current (sensitivity) and the peak shape (resolution, baseline, width), the optimum conditions for frequency, step potential, and amplitude have been found to be 200 ~ 400 Hz, 10 ~ 20 mV, and 15 ~ 25 mV, respectively. Generally, the sensitivity of measurements increases, if a more negative deposition potential and a longer deposition time are used. However, it should be stressed that potentials more negative than -1.5 V will cause hydrogen evolution, and a too long deposition time may lead to the overload of the electrodes with metal ions. In both cases, the measurements would be interfered. For simultaneous determination of the trace concentrations of Pb and Cd in a 0.02 M acetate buffer solution (pH = 4.5), the optimum potential is between -1.2 V and -1.4 V. The effect of different electrolytes on the SWASV measurement has been studied using 0.02 M acetate buffer (pH = 4.5), 0.1 M $NaNO_3$, 0.02 M acetate buffer+0.05 M $NaNO_3$, and 0.1 M HNO_3. The optimum supporting electrolyte has been found to be 0.02 M acetate buffer (pH = 4.5), which allows both high sensitivity and a long lifetime for the mercury film. The optimum pH value for the 0.02 M acetate buffer solution was found to be higher than 4. Medium exchange has been proved to evidently improve measurement sensitivity and shape of voltammogram. It is well known that dissolved oxygen is problematic in stripping voltammetry using macroelectrodes. However, no significant difference is seen in the SWASV measurements performed in the presence and the absence of dissolved oxygen using the Ir-UMEA sensors, which is in good agreement with the theory of microelectrodes. Acting as complexing agent or surfactant, organic matter has severe interferences on stripping voltammetric measurements, like reducing the reactivity of heavy metals, inhibiting the access of heavy metals to electrodes, and increasing the irreversibility of the ASV process. Therefore, for the determination of total concentrations of heavy metals in natural water, organic matter has to be destructed before analysis. In this study, river water samples are digested using both a self-made and a commercially available UV-digestion device. The SWASV measurements are not improved after the UV treatment with both of the apparatuses, which may in part be due to the influence of H_2O_2. However, it is found that trace concentrations of heavy metals (especially Pb, Cd) can be successfully determined without UV-digestion, if medium exchange and standard addition are used. Obvious interferences with intermetallic compounds, (e.g., Cu-Zn compounds) are not observed in the SWASV measurements, because the concentrations of heavy metals in the samples are low. By carrying out the measurements in synthetic solutions and in seawater, no formation of insoluble calomel (Hg_2Cl_2) on the electrode surfaces is observed, indicating that the UMEA sensors can be also used for seawater analysis.

Finally, the Ir-UMEA sensor is tested in determining the trace concentrations of heavy metals in river water samples. After on-site filtration and acidification, the samples are directly determined without UV-pretreatment, using medium exchange and standard addition in laboratory. The results obtained with the Ir-UMEA sensor show a reasonably good agreement with the HR-ICP-MS measurements (especially for Pb), indicating that the SWASV measurements are not significantly affected by organic matter and/or by the formation of intermetallic compounds. At more than half of the stations the concentrations of heavy metals measured using the UMEA sensor agree within ±20% with the results obtained with the HR-ICP-MS. The generally lower concentrations measured by the UMEA sensor are mostly because the measurements are performed without destruction of organic matter. The regional distributions of the heavy metal concentrations show the higher values in areas near large cities and with more industries.

In general, compared to the Pt- and Ir/Pt-UMEAs, the Ir-UMEAs have been shown satisfactory electrochemical and analytical behavior, indicating good design and fabrication quality. The coupling of the Ir-UMEAs and the square wave anodic stripping voltammetry has been proved to be a very powerful and promising analytical approach for the determination of trace concentrations of heavy metals in water. However, in order to get the pure properties of microelectrodes with the Ir-UMEAs, appropriate enlargement of the interelectrode distance (e.g., w/d ≥ 20) (LEE ET AL. 2001, SANDISON ET AL. 2002) is suggested for future work. By the combination of the UMEA sensor with an on-line UV-digestor or with an anti-fouling membrane (BELMONT-HEBERT ET AL. 1998), the successful development of a portable on-site or in situ trace metal analysis system in natural waters is very possible.

10. References

ABDELSALAM, M.E; DENUAULT, G; DANIELE, S. (2002): Calibrationless determination of cadmium, lead and copper in rain samples by stripping voltammetry at mercury microelectrodes effect of natural convection on the deposition step. Analytica Chimica Acta, 452: 65-75.

ACHTERBERG, E.P. & BRAUNGARDT, C. (1999): Stripping voltammetry for the determination of trace metals speciation and in-situ measurements of trace metal distributions in marine waters. Analytica Chimica Acta, 400: 381-397.

ACHTERBERG, E.P. & VAN DEN BERG, C.M.G. (1994): In-line ultraviolet- digestion of natural water samples for trace metal determination using an automated voltammetric system. Analytica Chmica Acta, 291:213-232.

ACHTERBERG, E.P; BRAUNGARDT, C.B; SANDFORD, R.C; WORSFOLD, P.J. (2001): UV digestion of seawater samples prior to the determination of copper using flow injection with chemiluminescence detection. Analytica Chimica Acta, 440: 27-36.

ALFASSI, Z.B. (1994): Determination of trace elements. VCH Publisher, New York, 394-424.

ALLEN, L.A.; LAECH, J.J. & HOUK, R.S. (1997): Spatial location of the space charge effect in individual ion clouds using monodisperse dried microparticulate injection with a twin spectrometer. Anal. Chem., 69: 2384-2391.

ALLOWAY, B.J. (1996): Schadstoff in der Umwelt. Spektrum Akademischer, Heidelberg.

AMATORE, C. IN RUBINSTEIN, I. (EDS.) (1995): Physical electrochemistry – Principles, methods, and applications. Marcel Dekker, New York, 131-208.

AMATORE, C.A. &. FOSSET, B. (1992): Space variables well fitted the study of steady-state and near steady-state diffusion at microdisk. J. Electroanal. Chem., 328: 21-32.

ANDERSON, J.L; BOWDEN, E.F. (1996): Dynamic electrochemistry: methodology and application. Anal. Chem., 68: 379R-444R.

ANDREWS, M.K. & HARRIS, P.D. (1998): Fabrication and sensing applications of microelectrodes on silicon substrates. Electroanalysis, 10: 1112-1118.

Aoki, K. & Osteryoung, J. (1981): Diffusion controlled current at the stationary finite disk electrode – Theory. J. Electroanal. Chem., 122: 19-35.

AOKI, K. (1993): Theory of ultramicroelectrodes. Electroanalysis, 5: 627-639.

BALDO, M.A; DANIELE, S; CORBETTA, M. & MAZZOCCHIN, G.A. (1995): Performance of platinum-based spherical mercury microelectrodes in cyclic voltammetry and stripping analysis. Electroanalysis, 7: 980-986.

BARANSKI, A.S. (1987): Rapid anodic stripping analysis with ultramicroelectrodes. Anal. Chem., 59:662-666.

BARD, A.J. & FAULKNER, L.R. (2001): Electrochemical methods – fundamentals and applications, 2rd Edi., John Wiley & Sons, New York, 231pp.

BARTLETT, P.N. & TAYLOR, S.L. (1998): An accurate microdisc simulation model for recessed microdisc electrodes. J. Electroanal. Chem., 453: 49-60.

BATLEY, G.E. (1986): Interferences in the determination of copper in natural waters by anodic stripping voltammetry. Analytica Chimica Acta, 189: 371-377.

BELMONT, C.; TERCIER, M.L.; BUFFLE, J.; FIACCABRINO, C.C. & KOUDELKA-HEP, M. (1996): Mercury-plated iridium-based microelectrode arrays for trace metals detection by voltammetry: optimum conditions and reliability. Analytica Chimica Acta, 329: 203-214.

BELMONT-HEBERT, C.; TERCIER; M.L.; BUFFLE, J.; FIACCABRINO, G.C.; ROOIJ, N.F & KOUDELKA-HEP, M. (1998): Gel-integrated microelectrode arrays for direct voltammetric measurements of heavy metals in natural waters and other complex media. Anal. Chem., 70: 2949-2956.

BERIET, C; FERRIGNO, R; GIRAULT, H.H. (2000): Simulation of the chronoamperometric response of a regular array of micro-disc electrodes. Journal of Electroanalytical Chemistry, 486: 56-64.

BOND, A.M.; LUSCOMBE, D.; OLDHAM, K.B. & ZOSKI, C.G. (1988): A comparison of the chronoamperometric response at inlaid and recessed disc microelectroes. J. Electroanal. Chem., 249: 1-14.

BOND, A.M.; OLDHAM, K.B. & ZOSKI, C.G. (1989): Steady-state voltammetry. Anal. Chim. Acta, 216: 177-230.

BOND, M.A. (1994): Past, present and future contribution of microelectrodes to analytical studies employing voltammetric detection, a review. Analyst, 119: R1-R20.

BRAININA, K.Z.; MALAKHOVA, N.A. & STOJKO, N.Y. (2000): Stripping voltammetry in environmental and food analysis. Fresenius J. Anal. Chem., 368: 307-325.

BRENDEL, D. & HOFFSTETTER-KUHN, S. (1996): Instrumentelle Analytik – Grundlagen, Geräte, Anwendungen (Übersetzung). Autoren: Skoog, D.A. & Leary, J.J., Springer-Verlag, Berlin, 577-609.

BRENNER, I.B. & TAYLOR, H.E. (1992): A critical review of inductively coupled plasma-mass spectrometry for geoanalysis, geochemistry, and hydrology. Part I: analytical performance. Critical Reviews in Analytical Chemistry, 23: 355-367.

BRETT, C.M.A. (1999): Electroanalytical techniques for the future: the challenges of miniaturization and of real-time measurements. Electroanalysis, 11: 1013-1016.

BRETT, C.M.A.; GARCIA, M.B.Q. & LIMA, J.L.F.C. (1996): On the suppression of Zinc-Copper interactions in square wave anodic stripping voltammetry in flowing solution by addition of gallium ions. Analytica Chimica Acta, 339: 167-172.

BRETT, C.M.A.; LIMA, J.F.C.; GARCIA, B.Q. (1994): Square-wave anodic stripping voltammetry in stationary and flowing solution: a comparative study. Analyst, 119: 1229-1233.

BREZONIK, P.L. (1976): Trace metal analysis by anodic stripping voltammetry: effect of sorption by natural and model organic compounds. Water Research, 10: 605-612.

BRUCKENSTEIN, S. (1987): Ohmic potential drop at electrodes exhibiting steady-state diffusion currents. Anal. Chem., 59: 2098-2101.

BUCHBERGER, W. (1998): Electrochemische Analyseverfahren – Grundlagen, Instrumentation, Anwendungen. Spektrum Akad. Verl., Heidelberg, 85-97.

BUFFLE, J.; TERCIER, M.L.; PARTHASARATHY, N. & WILKINSON, K.J. (1997): Analytical techniques for the in situ measurement and speciation of trace compounds in natural waters. Chimia, 51: 690-693.

BUFFLE, J.; WILKINSON, K.J.; TERCIER, M.-L. & PARTHASARATHY, N. (1997): In situ monitoring and speciation of trace metals in natural waters. Annali di Chimica, 87:67-83.

BUUFFLE, J. & HORVAI, H. (2000): In situ monitoring of aquatic systems – chemical analysis and speciation. John Wiley & Sons, Chichester.

CAI, X.; GLIDLE, A. & COOPER, J.M. (2000): Miniaturized electroanalytical sensor systems in micromachined structures. Electroanalysis,12: 631-639.

CAMMANN, K. (2001): Instrumentelle analytische Chemie – Verfahren, Anwendungen und Qualitätsicherung. Spektrum Akad. Verlag, Heidelberg, 7.1 - 7.63.

CAMMANN, K.; LEMKE, U;.ROHEN, J;.WILKEN, H. & WINTER, B.(1991): Chemo-und biosensoren-grundlagen und anwendungen. Angew. Chem.,103: 519-541.

CASSIDY, J; GHOROGHCHIAN, J; SARFARAZI, F; SMITH, J.J. &. PONS; S. (1986): Simulation of edge effects in electroanalytical experiments by orthogonal collocation-vi. Cyclic voltammetry at ultramicroelectrode ensembles. Electrochimica Acta, 31: 629-636.

CAUDILL, W.L; HOWELL, J.O. & WIGHTMAN, R.M. (1982): Flow rate independent amperometric cell. Anal. Chem., 54: 2532-2535.

CHENG, L.F. & MARTIN, C.R. (1988): Ultramicrodisk electrode ensembles prepared by incorporating carbon paste into a microporous host membrane. Anal. Chem., 60: 2163-2165.

CHING, S.; DUDEK, R. & TABET, E. (1994): Cyclic voltammetry with ultrmicroelectrdes. Journal of Chemical Education, 7: 602-605.

CHRISTIE, J.H; TURNER, J.A. & OSTERYOUNG, R.A. (1977): Square wave voltammetry at the dropping mercury electrode: theory. Analytical Chemistry, 49: 1899-1903.

CISZKOWSKA, M. & OSTERYOUNG, J.G. (1995): Voltammetry of metals at mercury film microelectrodes in the absence and the presence of varying concentrations of supporting electrolyte. Anal. Chem., 67: 1125-1131.

CISZKOWSKA, M. & STOJEK, Z. (1999): Voltammetry in solutions of low ionic strength. Electochemical and analytical aspects. Journal of Electroanalytical Chemistry, 466: 129-143.

CISZKOWSKA, M. & STOJEK, Z. (2000): Voltammetric and amperometric detection without added electrolyte, Anal. Chem., 72: 754A-760A.

COLOMBO, C. &. CONSTANT, M. G. VAN DEN BERG. (1998): In-line deoxygenation for flow analysis with voltammetric detection. Analytica Chimica Acta, 377: 229-240.

COPELAND, T.R; OSTERYOUNG, R.A. &. SKOGERBOE, R.K. (1974): Elimination of copper-zinc intermetallic interferences in anodic stripping voltammetry. Analytical Chemistry, 46: 2093-2097.

CROSMUN, S.T. &. DEAN, J.A. (1975): Pulsed anodic stripping voltammetry of zinc, cadmium and lead with a mercury-coated wax-impregnated graphite electrode. Analytica Chimica Acta, 75: 421-430.

DANIEL, S.; BALDO, M.A.; UGO, P. & MAZZOCHIN. G. (1989): Determination of heavy metals in real samples by anodic stripping voltammetry with mercury microelectrodes, Part 1: Application to wine. Analytica Chimica Acta, 319: 9-18.

DANIEL, S.; BALDO, M-A. UGO, P. & MAZZOCCHIN, A. (1989): Determination of heavy metals in real samples by anodic stripping voltammetry with mercury microelectrodes, Part 2: Application to rain and seawaters. Analytica. Chimica Acta, 219: 19-26.

DANIELE, S.; BALDO, M.A. & BRAGATO, C. (2002): In situ monitoring of electroactive species by using voltammetry at microelectrodes. J. Braz. Chem. Soc., 13: 425-432.

DANIELE, S.; BRAGATO, C.; BALDO, M.A.; WANG, J. & LU, J. (2000): The use of a remote stripping sensor for the determination of copper and mercury in the Lagoon of Venice. Analyst, 125: 731-735.

DATTA, M. & LANDOLT; D. (2000): Fundamental aspects and applications of electrochemical microfabrication. Electrochimica Acta., 45: 2535-2558.

DESLMONI, E.; PALMISANO, F. & SABBATINI, L. (1980): Simultaneous determination of tin and lead at the parts-per-billion level by coupling differential pulse anodic stripping voltammetry with a matrix exchange method. Anal. Chem., 52: 1889-1892.

DEUTSCHER, R.L. & FLETCHER, S. (1988): Nucleation on active sites. 4. Invention of an electronic method of counting the number of crystals as a function of time and the discovery of nucleation rate dispersion. J. Electroanal. Chem., 239: 17-54.

DIEDERICH, H-J.; MEYER, S. & SCHOTZ, F. (1994): Automatic adsorptive stripping voltammetry at thin- mercury film electrodes (TMFE). Fresenius J. Anal. Chem., 349: 670-675.

DONG, H. & KE, Y. (2000): Instrumental analysis (in Chinese). Chemical Industry Press, Beijing.

EMONS, H; BAADE, A. & SCHÖNING, M.J. (2000): Voltammetric determination of heavy metals in microvolumes of rainwater. Electroanalysis, 12: 1171-1175.

ESKLISSON, H.; HARALDSSON, C. & JAGNER, D. (1985): Determination of nickel and cobalt in natural waters and biological material by reductive chronopotentiometric stripping analysis in a flow system without sample deoxygenation. Analytica Chimica Acta, 175: 79-88.

FASSBENDER, F.; SCHMITT, G.; SCHOENING, M.J.; LUETH, H.; BUSS, G. & SCHULTZE, J.-W. (2000): Optimisation of passivation layers for corrosion protection of silicon-based microelectrode arrays. Sensors and Actuators B, 68: 128-133.

FAULKNER, L.R. & BARD, A.J. (1980): Electrochemical methods. John Wiley and Sons, New York, Chapter 6.

FEENEY, R. & KOUNAVES, S.P. (2000): Microfabricated ultramicroelectrode arrays: developments, advances, and application in environmental analysis, Electroanalysis, 12: 678-684.

FEENEY, R. & KOUNAVES, S.P. (2000): On-site analysis of arsenic in groundwater using a microfabricated gold ultramicroelectrode array. Anal. Chem., 72: 2222-2228.

FEENEY, R. & KOUNAVES, S.P. (2002): Voltammetric measurement of arsenic in natural waters. Talanta, 58: 23-31.

FEENEY, R.; HERDAN, J.; NOLAN, M.A.; TAN, S.H.; TARASOV, V.V. & KOUNAVES, S.P. (1998): Analytical characterization of microlithographically fabricated iridium-based ultramicroelectrode arrays. Electroanalysis, 10: 89-93.

FERRIGNO, R; BREVET, P.F. & GIRAULT, H.H. (1997): Finite element simulation of the chronoamperometric response of recessed and protruding microdisc electrodes. Electrochimica Acta, 42: 1895-1903.

FIACCABRINO, G.C. &. KOUDELKA-HEP, M. (1998): Thin-film microfabrication of electrochemical transducers. Electroanalysis, 10 (4): 217-222.

FIEDLER, H.J. & ROESLER, H.J. (1993): Spurenelemente in der Umwelt. Gustav Fischer Verlag, Jena.

FITCH, A. & EVANS, D.H. (1986): Use of microelectrodes for the study of a fast chemical step in an electrode reaction. J. Electroanal. Chem., 202: 83-92.

FLEISCHMANN, S.; PONS, ROLISON, D. & SCHMIDT, P.P. (1987): Ultramicroelectrodes. Datatech Science, Morganton, NC.

FLETCHER, S. & HORNE, M.D. (1999): Random assemblies of microelectrode (RAM electrodes) for electrochemical studies. Electrochemistry communications, 1: 502-512.

FLETCHER, S. (1994): Microelectrodes, Chemistry in Australia, February: 80-82.

FLORENCE, T.M. (1986): Electrochemical approaches to trace element speciation in waters a review. Analyst, 111: 489-503.

FORSTER, R.J. (1994): Microelectrodes: new dimensions in electrochemistry. Chemical Society Reviews, 289-297.

GAO, X.; LEE, J. & WHITE, H.S. (1995): Natural convection at microelectrodes. Anal Chem., 67: 1541-1545.

GAO, Y.; OSHITA, K.; LEE, K.H.; OSHIMA, M. & MOTOMIZU, S. (2002): Development of column-pretreatment chelating resins for matrix elimination/muti-element determination by inductively coupled plasma-mass spectrometry. Analyst, 127: 1713-1719.

GERLACH, R.W. &. KOWALSKI, B.R. (1982): The generalized standard addition method: intermetallic interferences in anodic stripping voltammetry. Analytica Chimica Acta, 134: 119-127.

GIL, E.P. & OSTAPEZUK, P. (1994): Potentiometric stripping determination of mercury (II) selenium (IV), copper (II) and lead (II) at a gold film electrode in water samples. Analytica Chimica Acta, 293: 55-65.

GOLIMOWSKI, J. & GOLIMOWSKA, K. (1996): Oxidation as pretreatment step in inorganic analysis of environmental samples. Analytica Chimica Acta, 325: 111-133.

GORSUCH, T.T. (1970): The destruction of organic matter, Pergamon, Oxford.

GOSSER, D.K.JR. (1993): Cyclic voltammetry – simulation and analysis of reaction mechanisms. VCH Publisher.

GRASSHOFF, K.; KREMLING, K. & EHRHARDT, M. (1999): Methods of seawater analysis. 3rd Edi., Wiley-Vch, Weinheim, 253-364.

GUENAT, O.T; FIACCABRINO, G.C; MORF, W.E; KOUDELKA-HEP, M. &. ROOIJ, N.F. (1999): Microfabricated chemical analysis systems for environmental applications. Chimia, 53: 87-90.

GUESHI, T.; TOKUDA, K. & MATSUDA, H. (1978): Voltemmetry at partially covered electrodes. 1. chronopotentiometry and chronoammperometry at model electrodes. J. Electroanal. Chem., 89: 247-260.

GUMINSKI, C. & GALUS, Z. (1986): Solubility data series – metals in mercury. Hirayama, C.; Guminski, C. & Galus, Z. (Eds.). Pergamon Press, Oxford, UK, Vol. 25.

HANSEN, P. (1991): Heavy metal determination using potentiometric stripping analysis. American Laboratory, April: 52-58.

HART, J.P. & WRING, S.A. (1997): Recent developments in the design and application of screen-printed electrochemical sensors for biomedical, environmental and industrial analyses. Trends in analytical chemistry, 16: 89-103.

HASWELL, J.S. (1997): Development and operating characteristics of micro flow injection analysis systems based on electroosmotic flow. Analyst, 122: 1R-10R.

HEINZE, J. (1981): Theory of cyclic voltammetry at micodisks electrodes. Ber. Bunsenges. Phys. Chem., 85: 1096-1103.

HEINZE, J. (1984): Cyclovoltammetrie - die "Spektroskopie" des Elektrochemikers. Angew. Chem., 96: 823-840.

HEINZE, J. (1993): Ultramicroelectrodes in electrochemistry. Angew. Chem. Int. Ed. Engl., 32: 1268-1288.

HENZE, G. & NEEB, R. (1986): Elektrochemische Analytik. Springer-Verlag, Berlin, 141pp.

HERDAN, J.; FEENEY, R. & KOUNAVES, S.P. (1998): Field evaluation of an electrochemical probe for in situ screening of heavy metals in groundwater. Environ. Sci. & Technol., 32: 131-136.

HILL, S.J. (1997): Speciation of trace metals in the environment. Chemical Society Reviews, 26: 291-298.

HOWELL, J.O. & WIGHTMAN, M. (1984): Ultrafast voltammetry and voltammetry in highly resistive solutions with microvoltammetric electrodes. Anal. Chem., 56: 524-529.

HOWELL, K.A.; ACHTERBERG, E.P.; BRAUNGARDT, C.B.; TAPPIN, A.D.; WORSFOLD, P.J. & TURNER, D.R. (2003): Voltemmetric in situ measurements of trace metals in coastal waters. Trends in analytical chemistry, 22: 828-835.

HOYER, B; FLORENCE, T.M. & BAILEY, G.E. (1987): Application of polymer-coated glassy carbon electrodes in anodic stripping voltammetry. Anal. Chem., 59: 1608-1614.

WANG J., IN: KISSINGER, P.T.; HEINEMAN, W.R. (EDS) (1996): Laboratory techniques in electroanalytical chemistry, Marcel Dekker, New York. pp. 719-738.

JAGNER, D. & SWEDEN, G. (1983): Computerised flow potentiometric stripping analysis. Trends in analytical chemistry, 2: 53-56.

JAGNER, D. (1982): Potentiometric stripping analysis. Analystica, 107: 593-599.

JAGNER, D; SAHLIN, E. & AXELSSON, B. (1993): Method for the determination of copper and lead (II) in tap water using a portable potentiometric stripping analyser. Analytica Chimica Acta, 278: 237-242.

JAGNER, D; SAHLIN, E. & RENMAN, L. (1996): Apparent formation of an oxidant by electrochemical reduction in the mercury (O, I, II) chloride system. Anal. Chem., 68: 1616-1622.

JOY, C.A. & PAWLEY, J.B. (1992): High-resolution scanning electron microscopy. Ultramicroscopy, 47: 80-100.

JU, H.; CHEN, H. & GAO, H. (1992): Investigation on microelectrodes. J Electronal Chem., 341: 35-46.

Kakihana, M.; Ikeuchi, H.; Sato, G.P. & Tokuda, K. (1981): Diffusion current at microdisk electrodes – application to accurate measurement of diffusion coefficients. J. Electroanal. Chem., 117: 201-211.

KALVODA, R. (1994): Review of adsorptive stripping voltammetry- assessment and prospects. Fresenius J. Anal. Chem., 349: 565-570.

KELLNER, R.; MERMET, J.-M.; OTTO, M. & WIDMER, H.M. (1998): Analytical chemistry. Wiley-Vch, Weinheim.

KHAKANI, M.A.E.; CHAKER, M. & DROGOFF, B.L. (1998): Iridium thin films deposited by radio-frequency megnetron sputtering. J. Vac. Sci. Technol., A, 16: 885-888.

KISSINGER, P.T.; & HEINEMAN, W.R. (1984): Laboratory techniques in electroanalytical chemistry, Marcel Dekker, New York.

KOLB, M.; RACH, P.; SCHÄFER, J. & WILD, A. (1992): Investigations of oxidative UV photolysis. Fresenius J. Anal. Chem., 342: 341-349.

KOMORSKY-LOVRIC, S. & BRANICA, M. 1992: Comparison of potentiometric stripping analysis and square-wave voltammetry with respect to the influence of triton X-100. Analytica Chimica Acta, 276: 361-366.

KOUNAVES, S.P. & BUFFLE, J. (1986): Deposition and stripping properties of mercury on iridium electrodes. J. Electrochem. Soc., 133: 2495-2497.

KOUNAVES, S.P. & BUFFLE, J. (1988): An iridium based mercury film electrode. Part II: comparison of mercury-film electrode behaviours: theory versus reality. J. Electroanal. Chem., 239: 113-123.

KOUNAVES, S.P. & DENG, W. (1991): An iridium based mercury ultramicroelectrode fabrication and characterization. J. Electroanal. Chem., 301: 77-85.

KOUNAVES, S.P. & DENG, W. (1993): Analytical utility of the iridium-based mercury ultramicroelectrode with square-wave anodic stripping voltammetry. Anal. Chem., 65: 375-379.

KOUNAVES, S.P.& BUFFLE, J. (1987): An iridium- based mercury-film electrode. J. Electronal. Chem., 216: 53-69.

KOUNAVES, S.P.; BUEHLER, M.G.; HECHT, M.H. & WEST, S. (2002): Chapter 16 Determination of Geochemistry on Mars Using an Array of Electrochemical Sensors. American Chemical Society, Washington, DC, pp. 306-319.

KOUNAVES, S.P.; DENG, W. & HALLOCK, P.R. (1994): Iridium-based ultramicroelectrode array fabricated by microlithography. Anal. Chem., 66: 418-423.

KOUNAVES, S.P; O'DEA, J.J.; CHANDRESEKHAR, P. & OSTERYOUNG, J. (1987): Square wave anodic stripping voltammetry at the mercury film electrode: theoretical treatment. Anal. Chem., 59: 386-289.

LAFLEUR, R.D; MYLAND, J.C. &. OLDBAM, K.B. (1990): Analytical microelectrode voltammetry with minimal instrumentation. Electroanalysis, 2: 223-228.

LEE, H.J.; BERIET, C.; FERRIGNO, R. & GIRAULT, H.H. (2001): Cyclic voltmmetry at a regular microdisc electrode array. J. Electroanal. Chem., 502: 138-145.

LINDEMANN, J. & LANDSBERG, R. (1971): Potentiostatic start measurements on partially blocked model electrodes. J. Electroanal. Chem., 30: 79.

MARQUES, A.L.B. &. CHIERICE, G.O. (1991): Elimination of the copper-zinc interference in anodic stripping voltammetry by addition of a complexing agent. Talanta, 38: 735-739.

MART, L. (1982): Minimization of accuracy risks in voltammetric ultratrace determination of heavy metals in natural waters. Talenta, 29: 1035-1040.

METROHM (1994): Gebrauchsanweisung der 693 VA-Processor und 694 VA-Stand. 7:35-41.

MEYER, H.; NAENDORF, B.; WITTKAMPF, M.; GRUENDIG, B.; CAMMANN, K.; KAKEROW, R.; MANOLI, Y.; MOKWA, W. &. ROSPERT, M. (1995): Microelectrode arrays as transducers for microanalysis systems, in van den Berg, A. & Bergveld, P. (Eds.), Micro Total analysis systems, Kluwer, Dordrecht, 245-248.

Microdisc array electrode (1992), Ecosse Sensors, Edinburgh.

MILLERO, F. & PIERROT, D. (2002): Speciation of metals in natural waters. Chemistry of marine water and sediments. International school on marine chemistry, 3rd, Palermo, Italy, Meeting Data Sept. 2000, 193-220.

MONK, P. (2001): Electroanalytical chemistry. John Wiley & Sons, New York, 158pp.

MONTENEGRO, M.I.; QUEIROS, M.A. & DASCHBACH, J.L. (1991): Microelectrodes: theory and applications. NATO ASI series. Kluwer Academic Publishers, Dordrecht, 3-16.

MOORE, J.W. & RAMAMOORTHY, S. (1984): Heavy metals in natural waters - applied monitoring and impact assessment. Springer-Verlag, New York.

MORF, W.E. & ROOIJ, N.F.D. (1997): Performance of amperometric sensors based on multiple microelectrode arrays. Sensors and Actuators B, 44: 538-541.

MORF, W.E. (1996): Theoretical treatment of the amperometric current response of multiple microelectrode arrays. Analytica Chimica Acta, 330: 139-149.

MORF, W.E. (1997): Theoretical treatment of the current vs. time response of microelectrode arrays to changes of potential, concentration, or flow. Analytica Chimica Acta, 341:121-127.

NASCIMENTO, V.B.; AUGELLI, M.A.; PEDROTTI, J.J.; GUTZ, I.G.R. & ANGNES, L. (1997): Arrays of microelectodes made from split integrated circuit chips. Electroanalysis, 9 (4): 335-339.

NEIMAN, E.Y; PETROVA, J.G; IGNATOV, V.I. &. DOLGOPOLOVA, G.M. (1980): The third element effect in anodic stripping voltammetry. Analytica Chimica Acta, 113: 277-285.

NOEL, S; TERCIER-WAEBER, M.L; LIN, L. & BUFFLE, J. (2003): Complexing gel integrated microelectrode arrays for direct detection of free metal ion concentrations in natural waters. J. Phys. IV France., 107: 965-968.

NOLAN, M.A. & KOUNAVES, S.P. (1998): Effects of mercury electrodeposition on the surface degradation of microlithographically fabricated iridium ultramicroelectrodes. Journal of Electroanalytical Chemistry, 453: 39-48.

NOLAN, M.A. & KOUNAVES, S.P. (1998): Failure analysis of microfabricated iridium ultramicroelectrodes in chloride media. Sensor and Actuators B, 50: 117-124.

NOLAN, M.A. & KOUNAVES, S.P. (1999): Effects of chloride ion concentration on mercury (I) chloride formation during ex situ and in situ mercury deposition with selected electrode substrates and electrodes. Anal. Chem., 71: 1176-1182.

NOLAN, M.A. & KOUNAVES, S.P. (1999): Microfabricated array of iridium microdisks as a substrate for direct determination of Cu^{2+} or Hg^{2+} using square-wave anodic stripping voltammetry. Anal. Chem., 71: 3567-3573.

NOLAN, M.A. & KOUNAVES, S.P. (2000): The source of the anomalous cathodic peak during ASV with in situ mercury film formation in chloride solutions. Electroanalysis, 12: 96-99.

NURNBERG, H.W. (1984): Trace analytical procedures with modern voltammetric determination methods for the investigation and monitoring of ecotoxic heavy-metals in natural waters and atmospheric precipitates. Sci. Tot. Environ., 37: 9-34.

Oldham, K.B. (1981): Edge effects in semi-infinite diffusion. J. Electroanal. Chem., 122: 1-17.

OSTAPCZUK, P. (1993): Present potentials and limitations in the determination of trace elements by potentiometric stripping analysis. Analytica Chimica Acta, 273: 35-40.

OSTERYOUNG, J.G. & OSTERYOUNG, R.A. (1985): Square wave voltammetry. Analytical Chemistry, 101A-110A.

PALCHETTI, I.; CAGNINI, A.; MASCINI, M. & TURNER, A.P.F. (1999): Characterisation of screen-printed electrodes for detection of heavy metals. Mikrochim Acta., 131: 65-73.

PALCHETTI, I.; MARRAZZA, GI. & MASCHINI, M. (2001): New procedures to obtain electrochemical sensors for heavy metal determination. Analytical Letter, 34: 813-824.

PEI, J.; TERCIER-WAEBER, M.L. & BUFFLE, J. (2000): Simultaneous determination and speciation of Zinc, Cadmium, Lead and Copper in natural water with minimum handling and artifacts, by voltammetry on a gel-integrated microelectrode array. Anal. Chem., 72: 161-171.

PENNER, R.M. & MARTIN, C.R.(1987): Preparation and electrochemical characterization of ultramicroelectrode ensembles. Anal. Chem., 59: 2625-2630.

PICCARDI, G.; UDISTI, R. (1987): Intermetallic Compounds and the determination of copper and zinc by anodic stripping voltammetry. Analytica Chimica Acta, 202: 151-157.

PISCH, J.; SCHÄFER, J.; FILDERSTADT,; FRAHNE, D. & REUTLINGEN. (1993): Voltammetrische Schwermetallbestimmungen in organisch hoch belasteten Flüssigkeiten nach UV-Aufschluß.GIT Fachz. Lab. 6/93: 500-505.

PONS, S.; FLEISCHMANN, M. (1987): The behavior of microelectrodes. Analytical Chemistry, 59: 1391-1399.

PRODENZIATI, M. (1994): Thick film sensors. In: Handbook of sensors and actuators, Vol. 1, Elsevier, Amsterdam.

REAY, R.J.; FLANNERY, A.F.; STORMENT, C.W.; KOUNAVES, S.P. & KOVACS, G.T.A: (1996): Microfabricated electrochemical analysis system for heavy metal detection. Sensors and Actuators B, 34: 450-455.

RELLER, H.; KIROWA-EISNER, E. & GILEDI, E. (1984): Ensembles of microelectrodes: digital simulation by the two-dimensional expanding grid method cyclic voltammetry, iR effects and applications. J. Electroanal. Chem., 161: 247-268.

RELLER, H.; KIROWA-EISNER, E.; GILEADI, E. (1982): Ensembles of microelectrodes a digital simulation. J. Electroanal. Chem., 138: 65-77.

RILEY, J.P. (1989): Chemical oceanography, 9: 197-245. Academic Press, London.

RISO, R.D.; CORRE, P.L. & CHAUMERY, C.J. (1997): Rapid and simultaneous analysis of trace metals (Cu, Pb and Cd) in seawater by potentiometric stripping analysis. Analitica Chimica Acta, 351:83-89.

RYAN, M.D; BOWDEN, E.F; CHAMBERS, J.Q. (1994): Dynamic electrochemistry: methodology and application. Anal. Chem., 66: 360R-427R.

SAGBERG, P. & LUND, W. (1982): Trace metal analysis by anodic-stripping voltammetry. Talanta, 29: 457-460.

SAMUEL, B; KOUNAVES, P. (1990): Iridium based ultramicroelectrodes, development and use in electrochemical analysis. Platinum Metals Rev., 34: 131-134.

SANDISON, M.E.; ANICET, N.; GLIDLE, A. & COOPER. J.M. (2002): Optimization of the geometry and porosity of microelectrode arrays for sensor design. Anal. Chem., 74: 5717-5725.

SAUR, D. IN: BRUTTEL, P.A. & SCHAFER, J. (Eds.) (1992): Samples preparation techniques in voltammetric trace analysis, Metrohm, Herisau, Switzerland.

SAUR, D.; SPAHN, E. & MAINZ. (1994): Die UV-Photolyse- ein nahezu reagenzienfreies Aufschlußverfahren für die Spurenanalytik. GIT Fachz. Lab., 2/94:104-106.

SCHARIFKER (1991): Ensembles of microelectrodes. In: MONTENEGRO, M.I.; QUEIROS, M.A. & DASCHBACH, J.L. (1991): Microelectrodes: theroy and applications. NATO ASI series. Kluwer Academic Publishers, Dordrecht, 228-240.

SCHARIFKER, B.R. (1988): Diffusion to ensembles of microelectrodes. J. Electroanal. Chem., 240: 61-76.

SCHMITT, G.; FAßBENDER, F.; LÜTH, H.; SCHÖNING, M.J.; SCHULTZE, J.W. & BUß, G. (2000): Passivation and corrosion of microelectrode arrays. Materials and Corrosion, 51: 20-25.

SCHMITT, G.; SCHULTZE, J.W.; FASSBENDER, F.; BUSS, G.; LÜTH, H. & SCHÖNING, M.J. (1999): passivaion and corrosion of microelectrode arrays. Electrochimica Acta, 44: 3865-3883.

SCHULTZE, J.W. & BRESSEL, A. (2001): Principles of electrochemical micro- and nano-system technologies. Electrochimica Acta, 47: 3-21.

SEDDON, B.J; SHAO, Y. &. GIRAULT, H.H. (1994): Printed microelectrode array and amperometric sensor for environmental monitoring. Electrochimica Acta, 39: 2377-2386.

SEDDON, B.J; SHAO, Y; FOST, J. & GIRAULT, H.H. (1994): The application of excimer laser micromachining for the fabrication of disc microelectrodes. Electrochimica Acta., 39: 783-791.

SHOUP, D. & SZABO, A. (1984): Chronoamperometry at an ensemble of microdisk electrodes. J. Electroanal. Chem., 160: 19-26.

SHUMAN, M.S. &. WOODWARD, G.P; JR. (1976): Intermetallic compound formation between copper and zinc in mercury and its effects on anodic stripping voltammetry. Analytical Chemistry, 48: 1979-1983.

SILVA, P.R.M.; KHAKANI, M.A.E.L.; CHAKER, M.; DUFRESNE, A. &. COURCHESNE, F. (2001): Simultaneous determination of Cd, Pb, and Cu metal trace concentrations in water certified samples and soil extracts by means of Hg-electroplated Ir microelectrode array based sensors. Sensors and Actuators B, 76: 250-257.

SILVA, P.R.M.; KHAKANI, M.A.EL.; CHAKER, M.; CHAMPAGNE, G.Y.; CHEVALET, J.; GASTONGUAY, L.; LACASSE, R. &. LADOUCEUR, M. (1999): Development of Hg-electroplated-iridium based microelectrode arrays for heavy metal traces analysis. Analytica Chimica Acta, 385: 249-255.

SILVA, P.R.M.; KHAKANII, M.A.EL.; DROGOFF, B.LE.; CHAKER, M. & VIJH, A.K. (1999): Mercury-electroplated-iridium microelectrode array based sensors for detection of heavy metal ultratraces: optimisation of the mercury charge. Sensors and Actuators B, 60: 161-167.

SOARES, H.M.V.M.; TERESA, M. & VASCONCELOS, S.D. (1995): Potentiometric stripping analysis vs. Differential pulse anodic stripping voltammetry for copper (II) analysis at relatively positive deposition. Analytica Chemica Acta, 303: 255-263.

SOTTERY, J.P. & ANDERSON C.W. (1987): Short-pulse rapid-scan stripping voltammetry at a thin mercury film carbon fiber electrode Anal. Chem., 59: 140-144.

STATHAM, P.J. (1985): The determination of dissolved manganese and cadmium in sea water at low nmol 1-1 concentrations by chelation and extraction flowed by electrothermal atomic absorption spectrometry. Analytica Chimica Acta, 169: 149-159.

STOJEK, Z. (1991): New possibilities in analytical chemistry connected with voltammetric applications of microelectrodes. Mikrochim. Acta [Wien], 11: 353-361.

STULIK, K. (1992): Activation of solid electrodes. Electroanalysis, 4: 829-834.

STULIK, K.; AMATORE, C.; HOLUB, K.; MARECEEK, & KUTNER, W. (2000): Microelectrodes. Definitions, characterization, and applications. Pure Appl. Chem., 72: 1483-1492.

STUMM, W. & MORGAN, J.J. (1996): Aquatic chemistry – chemical equilibria and rates in natural waters. John Wiley & Sons, 3rd Edi., New York.

SUZUKI, H. (2000): Advances in the microfabrication of electrochemical sensors. Electroanalysis, 12: 703-715.

SWAN, P.N. (1980): Dissertation, University of Southampton.

TAILLEFERT, M; LUTHER III, G.W. &. NUZZIO, D.B. (2000): The application of electrochemical tools for in situ measurements in aquatic systems. Electroanalysis, 12(6): 401-412.

TALLMAN, D.E.; PETERSEN S.L. (1990): Composite electrodes for electroanalysis: principles and Applications. Electroanalysis, 2: 499-510.

TAN, S.H. & KOUNAVES, S.P. (1998): Determination of selenium (IV) at a microfabricated gold ultramicroelectrode array using square wave anodic stripping voltammetry. Electroanalysis, 10: 364-368.

TERCIER, M.L. & BUFFLE, J. (1993): In situ voltammetric measurements in natural waters: future prospects and challenges. Electroanalysis, 5: 187-200.

TERCIER, M.L. &. BUFFLE, J. (1996): Antifouling membrane-covered voltammetric microsensor for in situ measurements in natural waters. Anal. Chem., 68: 3670-3678.

TERCIER, M.L.; PARTHASARATHY, N. & BUFFLE, J. (1995): Reproducible, reliable and rugged Hg-plated Ir-based microelectrode for in situ measurements in natural waters. Electroanalysis, 7: 55-63.

TERCIER, M-L.; BUFFLE, J.; GRAZIOTTIN, F. & KOUDELKA-HEP, M. (1999): Novel voltammetric probe for in-situ trace element monitoring. Sea technology, 40 (5): 74-79.

TERCIER-WAEBER, M. -L. & BUFFLE, J. (2000): Submersible online oxygen removal system coupled to an in situ voltammetric probe for trace element monitoring in freshwater. Environ. Sci. Technol., 34: 4018-4024.

TERCIER-WAEBER, M.L.; BELMONT-HEBERT, C. & BUFFLE, J. (1998): Real-time continuous Mn (II) monitoring in lakes using a novel voltammetric in situ profiling system. Environ. Sci. Technol., 32: 1515-1521.

TERCIER-WAEBER, M.L.; BUFFLE, J.; CONFALONIERI, F.; RICCARDI, G.; SINA, A.; GRAZIOTTIN, F.; FIACCABRINO, G.C. & KOUDELKA-HEP, M.K. (1999): Submersible voltammetrc probes for in situ real-time trace element measurements in surface water, groundwater and sediment-water interface. Meas. Sci. Technol., 10: 1202-1213.

TESSIER, A. & TURNER, D.R. (1995): Metal speciation and bioavailability in aquatic systems. John Wiley & Sons, Chichester.

UHLIG, A.; PAESCHKE, M.; SCHNAKENBERG, U.; HINTSCHE, R.; DIEDERICH, H.J. & SCHOLZ, F. (1995): Chip-array electrodes for simultaneous stripping analysis of trace metals. Sensors and Actuators B, 25: 899-903.

UHLIG, A.; SCHNAKENBERG, U. HINTSCHE, R. (1997): Highly sensitive heavy metal analysis on platinum- and gold-Ultramicroelectrode arrays. Electroanalysis, 9: 125-129.

VAN DEN BERG, C.M.G. (1989): The electroanalytical chemistry of seawater. In: Chemical Oceanography, J.P. Riley (Ed.), Academic Press, London, Vol. 9, 197-245.

VITRE, R.R.D.; TERCIER, M.-L. & BUFFLE, J. (1991): In situ voltammetric measurements in natural waters: the advantages of microelectrodes. Analytical Proceedings, 28: 74-75.

WAHDAT, F. &. NEEB, R. (1987): Die aufschlussfreie Bestimmung von Schwermetallen (Zn, Cd, Pb, Cu) in Getränken (Wein) durch inverse Voltammetrie in einer Durchflusszelle. Fresenius Z. Chem., 327: 175-178.

WANG, J. & LUOT, D.B. (1984): Effect of surface-active compounds on voltammetric stripping analysis at the mercury film electrode. Talanta, 31: 703-707.

WANG, J. & TIAN, B. (1992): Screen-printed stripping voltammetric/potentiometric electrode for decentralized testing of trace lead. Anal. Chem., 64: 1706-1709.

WANG, J. &. ARMALIS, S. (1995): Stripping voltammetry at microdisk composite electrode assembly. Electroanalysis, 7: 958-961.

WANG, J. &. GREENE, B. (1983): Stripping analysis of zinc in natural waters utilizing the medium exchange method. Water. Res., 17: 1635-1638.

WANG, J. (1985): Stripping analysis – principles, instrumentation, and applications. VCH publisher, Deerfield Beach, Florida.

WANG, J. (2000): Analytical electrochemistry. 2nd Edi., A John Wiley & Sons, New York, 75pp.

WANG, J. (2000): In situ electrochemical monitoring: from remote sensors to submersible microlaboratories. LRA, 12: 178-182.

WANG, J. LU, J; TIAN, B. & YARNITZKY, C. (1993): Screen-printed ultramicroelectrode arrays for on-site stripping measurements of trace metals. Journal of Electroanalytical Chemistry, 361: 77-83.

WANG, J.; ADENIYI, W.K. & KOUNAVES, S.P. (1999): Adsorptive stripping analysis of trace nickel at iridium-based ultramicroelectrode arrays. Electroanalysis, 12: 44-47.

WANG, J.; BRENNSTEINER, A. & SYLWESTER, A.P. (1990): Composite electrodes based on carbonized poly (acrylonitrile) foams. Anal. Chem., 62: 1102-1104.

WANG, J.; FOSTER, N.; ARMALIS, S.; LARSON, D.; ZIRINO, A. & OLSEN, K. (1995): Remote stripping electrode for in situ monitoring of labile copper in the marine environment, Analytica Chimica Acta, 310: 223-231.

WANG, J.; LARSON, D.; FOSTER, N.; ARMALIS, S.; LU, J. & RONGRONG, X. (1995): Remote electrochemical sensor for trace metal contaminants, Anal. Chem., 67: 1481-1485.

WANG, J.; LU, J.; MACDONALD, D. & AUGELLI, M.A. (1999): In-situ flow probe for improving the performance of electrochemical stripping analysis. Fresenius J. Anal. Chem., 364: 28-31.

WANG, J.; LU, J.M.; TIAN, B.M. & YARNITZKY, C. (1993): Screen-printed ultramicroelectrode arrays for on-site stripping measurements of trace metals. J. Electroanal. Chem., 361: 77-83.

WANG, J.; TIAN, B.; LU, J,; WANG, J.; LUO, D. & MACDONALD, D.(1998): Remote electrochemical sensor for monitoring trace mercury. Electroanalysis, 10: 399-402.

WANG, J.; TLAN, B.; WANG, J.; LU, J.; OLSEN, C.; YARNITZKY, C.; OLSEN, K.; HAMMERSTROM, D. & BENNETT, W. (1999): Stripping analysis into the 21st century: faster, smaller, cheaper, simpler and better. Analytica Chimica Acta, 385: 429-435.

WANG, J; FARIAS, P.A.M. & LUO, D-B. (1984): Improved resolution in stripping analysis using the formation of intermetallic compounds. Anal. Chem., 56: 2379-2382.

WANG, J; PEDRERO, M; SAKSIUND, H; HAMMERICH, O. & PINGARRON, J. (1996): Electrochemical activation of screen-printed carbon strips. Analyst. March., 121: 345-350.

WEBER, S.G. (1989): Signal-to-noise ratio in microelectrode-array-based electrochemical detectors. Anal. Chem., 61: 295-302.

WECHTER, C. & OSTERYOUNG, J. (1989): Square wave and linear scan anodic stripping voltammetry at iridium-based mercury film electrodes. Anal. Chem., 61:2092-2097.

WECHTER, C. & OSTERYOUNG, J. (1990): Voltammetric characterization of small platinum-iridium-based mercury film electrodes. Analytica Chimica Acta, 234: 275-284.

WEHMEYER, K.R.; DEAKIN, M.R.& WIGHTMAN, M. (1985): Eletroanalytical properties of band electrodes of submicrometer width. Anal. Chem., 57: 1913-1916.

WEISSHAAR, D.E. & TALLMAN, D.E. (1983): Chronoamperometric response at carbon-Based composite electrodes. Anal. Chem., 55: 1146-1151.

WEST, A.C. (1993): Ohmic interactions within electrode ensembles. J. Electrochem. Soc., 140: 134-139.

WIGHTMAN, R.M. & WIPF, D.O. (1989): voltmmetry at ultrmicroelectrodes, in Bard, A.J. (ed.), Electroanalytical chemistry, Vol. 15, Marcel Dekker, New York.

WIGHTMAN, R.M. (1981): Microvoltammetric Electrodes. Anal. Chem., 53: 1125A-1134A.

WIGHTMAN, R.M. (1988): Voltammetry with Microscopic Electrodes in New Domains. Science, 240: 415-420.

WOJCIECHOWSKI, M. & BALCERZAK, J. (1990): Microvolume electrochemical cell employing cylindrical graphite fiber microelectrodes. Analitica Chimica Acta., 237: 127-133.

WOJCIECHOWSKI, M. & BALCERZAK, J. (1991): Square-wave anodic stripping voltammetry of lead and cadmium at cylindrical graphite fiber microelectrodes with in situ plated mercury films. Analytica Chimica Acta, 249: 433-445.

WOJCIECHOWSKI, M.; BALCERZAK, J. (1990): Square-wave anodic stripping voltammetry at glassy-carbon-based thin mercury film electrodes in solutions containing dissolved oxygen. Anal. Chem., 62: 1325-1331.

WOLLENBERGER, U.; HINTSCHE, R. & SCHELLER, F. (1995): Biosensors for analytical mic.rosystems, Microsyst Technol., 1: 75.

WONG, D.K.Y. & EWING, A.G. (1990): Anodic stripping voltammetry at mercury films deposited on ultrasmall carbon-ring electrodes. Anal. Chem., 62: 2697-2702.

WRING, S.A. & HART, J.P. (1992): Chemically modified, screen-printed carbon electrodes. Analyst, 117: 1281-1286.

WU, H.P. (1993): Fabrication and characterization of a new class of microelectrode arrays exhibiting steady-state current behavior. Anal. Chem., 65: 1643-1646.

WU, H.P. (1994): Nature and stability of mercury thin films on glassy carbon electrodes under fast-scan anodic stripping voltammetry. Anal. Chem., 66: 3151-3157.

XIE, X.; STUEBEN, D.; BERNER, Z.; ALBERS, J.; HINTSCHE, R. & JANTZEN, E. (2004): Development of an ultramicroelectrode arrays (UMEAs) sensor for trace heavy metal measurement in water. Sensor and Actuators, B 97: 168-173.

YARNITZKY, C.; WANG, J. & TIAN, B. (2000): Hand-held lead analyzer. Talanta, 51: 333-338.

YOKOI, K.; TOMISAKI, T.; KOIDE, T. & VAN DEN BERG, M.G. (1995): Effective UV photolytic decomposition of organic compounds with a low- pressure mercury lamp as pre-treatment for voltammetric analysis of trace metals. Fresenius J. Anal. Chem., 352: 547-549.

YOKOI, K; KURATANI, Y. & KUBONO, K. (2002): Photolytic decomposition of saccharides with a high-output, low-pressure mercury lamp for voltammetric trace analysis. Analytica Sciences, 18: 1155-1157.

YOKOI, K; YAKUSHIJI, M;.HATANAKA, M; KUBONO, K; KOIDE, T. (1999): Novel photolytic decompositon method of organic compounds with a high output low-pressure mercury lamp for voltammetric trace metal analysis. Fresenius J. Anal. Chem., 365: 364-367.

ZUHRI, A. & VOELTER, W. (1998): Applications of adsorptive stripping voltammetry for the trace analysis of metals, pharmaceuticals and biomolecules. Fresenius J. Anal. Chem., 360: 1-9.

ZHANG, Z. (2000): Ultramicroelectrodes in electrochemistry (in Chinese). Science Press, Beijing, 341-376.

ZOSKI, C.G. (1990): A survey of steady-state microelectrodes and experimental approaches to a voltammetric steady state. J. Electroanal. Chem., 296: 317-333.

ZOSKI, C.G.; BOND, A.M.; ALLINSON, E.T. & OLDHAM, K.B. (1990): How long does it take a microelectrode to reach a voltammetric steady state? Anal. Chem., 62: 37-45.

Appendix 1: The reproducibility and accuracy of trace metal measurements using Ir-UMEA sensor

	Ir 27_plating 9		Ir 27_plating 5		Ir 24_plating 4		Ir 29_plating 1		Ir 29_plating 2	
No.	**Cd** (1 µg/l)	**Pb** (1 µg/l)	**Cd** (1 µg/l)	**Pb** (1 µg/l)	**Cd** (1 µg/l)	**Pb** (1 µg/l)	**Cd** (1 µg/l)	**Pb** (1 µg/l)	**Cd** (1 µg/l)	**Pb** (1 µg/l)
1	0.95	0.84	0.96	0.76	0.97	0.96	0.83	0.88	1.16	0.91
2	0.89	0.93	0.93	1.08	0.98	1.11	0.82	0.82	1.02	1.14
3	0.98	0.92	1.15	1.19	1.03	1.18	0.96	0.85	1.02	1.12
4	1.14	1.01	1.11	1.06	1.15	1.09	0.90	0.83	1.01	1.18
5	1.21	1.12	0.79	1.05	1.15	1.13	0.95	0.92	0.98	1.08
6	1.24	1.07	0.97	1.19	1.18	1.10	1.04	1.09		
7	1.05	1.17	1.01	1.15	1.19	1.18	1.12	1.11		
8	1.05	1.10	1.13	1.18	0.99	1.06	0.91	0.79		
9	1.19	1.15	1.11	1.18	0.98	1.05	0.98	0.91		
10	1.15	0.99	0.98	0.91	1.28	1.06				
11	1.03	1.11	0.92	1.14	0.99	1.14				
12	0.95	1.07	0.89	0.86	0.82	1.10				
13	0.86	0.88	0.93	0.86	0.81	0.92				
14	0.94	1.04	0.96	1.05	0.90	0.99				
15	0.95	1.04	0.96	1.10	0.87	1.03				
16	0.80	0.98	0.78	1.10	1.01	0.98				
17	1.03	1.04	1.04	1.16	1.02	1.00				
19	0.96	0.89	0.91	1.10	0.94	1.00				
20	0.89	0.86	1.17	0.87	0.88	0.99				
21	0.93	0.95	1.12	1.18	0.82	0.98				
22	0.93	0.88	0.92		0.87	0.85				
23	0.82	0.89	0.97		1.10	0.95				
24	0.83	0.90	0.99		0.97	1.12				
25	0.88	0.94	0.98		1.05	0.87				
26	0.81	0.82	0.91		1.07	0.85				
27	0.81	0.85	0.78		1.09	1.12				
28	0.86	0.88	0.83		1.10	1.02				
29	0.88	1.01	0.85		0.83	0.83				
30	0.86	0.90			0.88	0.99				
31	0.89	0.92			0.98	0.86				
32	0.93	0.95			0.84	1.11				
33	0.82	0.98			1.11	0.85				
34	0.84	1.05			0.97	0.84				
35	0.96	1.11			1.08	0.98				
36	0.80	0.87			1.19	0.93				
37	1.00	1.03			0.89	1.04				
38	0.93	1.01			1.00	0.92				
39	0.82	0.95			1.10	1.13				
40	0.85	0.95			1.10	1.11				
41	0.92	1.06			0.99	0.88				
42	0.99	1.11			1.10	1.27				
43	0.82	1.01			0.93	0.71				
44	0.97	1.06			0.96					
45	0.91	1.03			0.95					
46	1.06	1.15			0.92					
47	0.95	1.02								
48	1.03	1.11								
49	1.05	1.13								
50	0.89	0.93								

Appendix 2: The concentrations of trace metals in river water samples measured by Ir-UMEA sensor and HR-ICP-MS

Sample	Zn_ASV	Zn_ICP	Pb_ASV	Pb_ICP	Cu_ASV	Cu_ICP	Cd_ASV	Cd_ICP
p 1	11.97	13.30	0.85	1.01	-	2.68	-	0.050
p 2	4.92	8.91	0.23	0.25	0.57	0.63	-	0.015
p 3	8.15	9.08	0.38	0.45	0.95	1.12	-	0.024
p 4	8.31	9.40	0.08	0.07	0.64	0.26	-	0.014
p 5	-	7.61	-	0.22	-	1.15	-	0.583
p 6	2.04	3.97	0.11	0.09	0.53	1.31	-	0.020
p 7	7.89	10.61	0.09	0.11	1.97	2.41	0.035	0.049
p 8	9.96	8.92	0.09	0.08	0.26	0.32	0.018	0.023
p 9	5.14	5.00	0.09	0.08	0.41	0.47	-	0.014
p 10	2.23	2.80	0.04	0.07	0.45	0.46	-	0.010
p 11	1.25	2.06	0.03	0.04	0.14	0.16	-	0.007
p 12	15.52	23.93	0.12	0.06	2.03	1.92	-	0.023
p 13	2.79	3.77	0.08	0.12	0.40	0.41	-	0.011
p 14	5.72	12.63	0.13	0.10	1.40	2.13	0.019	0.028
p 15	4.73	7.61	0.17	0.28	0.61	0.45	0.025	0.032
p 16	2.86	3.14	0.05	0.06	0.05	0.07	-	0.011
p 17	1.65	4.20	0.09	0.07	0.96	1.02	-	0.012
p 18	4.89	4.70	0.15	0.13	0.61	0.71	-	0.012
p 19	4.79	5.01	0.14	0.13	1.88	1.37	-	0.016
p 20	3.85	3.62	0.07	0.05	0.99	0.78	-	0.013

- not available

Appendix 3: Parameters related to river water sampling

	Code	Site	River	Easting	Northing	DOC (mg/l)	t (°C)	pH	O_2 (mg/l)	Conductivity (µs/cm)
p 1	PF034	Russheim	Pfinz	3459887.00	5448881.00	3.5	7.8	8.17	10.1	742
p 2	SB014	Russheim	al	3458121.00	5449588.00	8	8	7.82	9.3	756
p 3	SB020	Philippsburg	Saalbach	3461974.00	5455253.00	6.3	10	8.07	9.4	990
p 4	SB025	Mündung-Wagbach	Wagbach	3463510.00	5461115.00	8.7	10.2	7.78	7.7	1061
p 5	SB033	Altlußheim	Kriegbach	3463323.00	5462063.00	6.1	7.7	8.29	10.1	1283
p 6	XX426	Mannheim, Rhein	Rhein	3460580.00	5483130.00	2.1	12	8.02	8.5	472
p 7	YY003	Mannheim, Neckar	Neckar	3461600.00	5484340.00	5.6	10.7	8.00	7.1	1015
p 8	EL840	Riegel	Dreisam	3407449.00	5335081.00	1.3	8	7.79	9.9	303
p 9	EL900	Riegel-2	Elz	3407543.00	5335940.00	0.9	8.2	7.85	10.8	257
p 10	XX225	Vogelgrün	Rhein	3393760.00	5321210.00	1.5	10.9	8.04	10.9	462
p 11	XX224	Breisach	Rhein	3394500.00	5321310.00	1.3	9.3	8.32	12.1	408
p 12	XL100	Jamborschwelle	al	3402463.00	5344943.00	5.5	12.2	7.47	8.4	850
p 13	EL904	Rust	Elz	3406285.00	5344184.00	0.8	8.7	7.88	10.07	287
p 14	EL911	Schutterkanal	Nonnenweier	3408339.00	5358208.00	7.6	12.1	8.06	9.1	1186
p 15	MU045	Steinmauern	Murg	3440785.00	5417367.00	4.1	8.1	7.48	10.5	216
p 16	AC029	Iffezheim	Sandbach	3435594.00	5410167.00	0.8	10.1	7.77	8.3	491
p 17	RE014	Helmlingen	Rench	3423791.00	5397655.00	0.4	8.5	7.66	9	287
p 18	KI901	Kehl	Kinzig	3413294.00	5384227.00	11.5	9	7.80	10.1	327
p 19	KI703	Offenburg	Kinzig	3422495.00	5368420.00	2.8	8	8.02	13	192
p 20	RE006	Oberkirch	Rench	3432718.00	5376615.00	1.1	8.6	8.60	12.6	157

Appendix 4: Different plating times used for a constant charge (chip 1)

	Plating 9	Plating 5		Plating 9	Plating 5
Time (s)	Charge (C)	Charge (C)	Time (s)	Charge (C)	Charge (C)
0.1	6.12E-08	9.69E-08	106.1	6.11E-05	6.83E-05
2.1	1.02E-06	1.19E-06	108.1	6.23E-05	6.97E-05
4.1	2.00E-06	2.27E-06	110.1	6.36E-05	7.11E-05
6.1	3.00E-06	3.37E-06	112.1	6.48E-05	7.25E-05
8.1	4.02E-06	4.48E-06	114.1	6.60E-05	7.39E-05
10.1	5.07E-06	5.60E-06	116.1	6.73E-05	7.53E-05
12.1	6.13E-06	6.74E-06	118.1	6.85E-05	7.68E-05
14.1	7.20E-06	7.89E-06	120.1	6.97E-05	7.82E-05
16.1	8.29E-06	9.05E-06	122.1	7.10E-05	7.96E-05
18.1	9.39E-06	1.02E-05	124.1	7.22E-05	8.11E-05
20.1	1.05E-05	1.14E-05	126.1	7.34E-05	8.25E-05
22.1	1.16E-05	1.26E-05	128.1	7.47E-05	8.39E-05
24.1	1.27E-05	1.38E-05	130.1	7.59E-05	8.54E-05
26.1	1.39E-05	1.50E-05	132.1	7.72E-05	8.68E-05
28.1	1.50E-05	1.63E-05	134.1	7.84E-05	8.83E-05
30.1	1.61E-05	1.75E-05	136.1	7.97E-05	8.98E-05
32.1	1.73E-05	1.87E-05	138.1	8.09E-05	9.12E-05
34.1	1.84E-05	2.00E-05	140.1	8.22E-05	9.27E-05
36.1	1.96E-05	2.12E-05	142.1	8.34E-05	9.42E-05
38.1	2.08E-05	2.25E-05	144.1	8.47E-05	9.57E-05
40.1	2.19E-05	2.38E-05	146.1	8.59E-05	9.72E-05
42.1	2.31E-05	2.50E-05	148.1	8.72E-05	9.87E-05
44.1	2.43E-05	2.63E-05	150.1	8.84E-05	1.00E-04
46.1	2.54E-05	2.76E-05	152.1	8.97E-05	1.02E-04
48.1	2.66E-05	2.89E-05	154.1	9.10E-05	1.03E-04
50.1	2.78E-05	3.02E-05	156.1	9.22E-05	1.05E-04
52.1	2.89E-05	3.15E-05	158.1	9.35E-05	1.06E-04
54.1	3.01E-05	3.28E-05	160.1	9.48E-05	1.08E-04
56.1	3.13E-05	3.42E-05	162.1	9.60E-05	1.09E-04
58.1	3.25E-05	3.55E-05	164.1	9.73E-05	1.11E-04
60.1	3.36E-05	3.68E-05	166.1	9.86E-05	1.12E-04
62.1	3.48E-05	3.82E-05	168.1	9.98E-05	1.14E-04
64.1	3.60E-05	3.95E-05	170.1	1.01E-04	1.15E-04
66.1	3.72E-05	4.08E-05	172.1	1.02E-04	1.17E-04
68.1	3.84E-05	4.22E-05	174.1	1.04E-04	1.18E-04
70.1	3.95E-05	4.35E-05	176.1	1.05E-04	1.20E-04
72.1	4.07E-05	4.49E-05	178.1	1.06E-04	1.21E-04
74.1	4.19E-05	4.63E-05	180.1	1.08E-04	1.23E-04
76.1	4.31E-05	4.76E-05	182.1	1.09E-04	1.25E-04
78.1	4.43E-05	4.90E-05	184.1	1.10E-04	1.26E-04
80.1	4.55E-05	5.03E-05	186.1	1.11E-04	1.28E-04
82.1	4.67E-05	5.17E-05	188.1	1.13E-04	1.29E-04
84.1	4.78E-05	5.31E-05	190.1	1.14E-04	1.31E-04
86.1	4.90E-05	5.44E-05	192.1	1.15E-04	1.32E-04
88.1	5.02E-05	5.58E-05	194.1	1.17E-04	1.34E-04
90.1	5.14E-05	5.72E-05	196.1	1.18E-04	1.36E-04
92.1	5.26E-05	5.86E-05	198.1	1.19E-04	1.37E-04
94.1	5.38E-05	5.99E-05	200.1	1.20E-04	1.39E-04
96.1	5.50E-05	6.13E-05	202.1	1.22E-04	1.40E-04
98.1	5.62E-05	6.27E-05	204.1	1.23E-04	1.42E-04
100.1	5.75E-05	6.41E-05	206.1	1.24E-04	1.43E-04
102.1	5.87E-05	6.55E-05	208.1	1.26E-04	1.45E-04
104.1	5.99E-05	6.69E-05	210.1	1.27E-04	1.47E-04

Continue

	Plating 9	Plating 5		Plating 9	Plating 5
Time (s)	Charge (C)	Charge (C)	Time (s)	Charge (C)	Charge (C)
212.1	1.28E-04	1.48E-04	320.1	1.99E-04	2.40E-04
214.1	1.29E-04	1.50E-04	322.1	2.00E-04	2.42E-04
216.1	1.31E-04	1.52E-04	324.1	2.01E-04	2.43E-04
218.1	1.32E-04	1.53E-04	326.1	2.03E-04	2.45E-04
220.1	1.33E-04	1.55E-04	328.1	2.04E-04	2.47E-04
222.1	1.35E-04	1.56E-04	330.1	2.05E-04	2.49E-04
224.1	1.36E-04	1.58E-04	332.1	2.07E-04	2.51E-04
226.1	1.37E-04	1.60E-04	334.1	2.08E-04	2.52E-04
228.1	1.39E-04	1.61E-04	336.1	2.09E-04	2.54E-04
230.1	1.40E-04	1.63E-04	338.1	2.10E-04	2.56E-04
232.1	1.41E-04	1.65E-04	340.1	2.12E-04	2.58E-04
234.1	1.42E-04	1.66E-04	342.1	2.13E-04	2.60E-04
236.1	1.44E-04	1.68E-04	344.1	2.14E-04	2.61E-04
238.1	1.45E-04	1.69E-04	346.1	2.16E-04	2.63E-04
240.1	1.46E-04	1.71E-04	348.1	2.17E-04	2.65E-04
242.1	1.48E-04	1.73E-04	350.1	2.18E-04	2.67E-04
244.1	1.49E-04	1.74E-04	352.1	2.20E-04	2.69E-04
246.1	1.50E-04	1.76E-04	354.1	2.21E-04	
248.1	1.52E-04	1.78E-04	356.1	2.22E-04	
250.1	1.53E-04	1.79E-04	358.1	2.24E-04	
252.1	1.54E-04	1.81E-04	360.1	2.25E-04	
254.1	1.55E-04	1.83E-04	362.1	2.26E-04	
256.1	1.57E-04	1.84E-04	364.1	2.28E-04	
258.1	1.58E-04	1.86E-04	366.1	2.29E-04	
260.1	1.59E-04	1.88E-04	368.1	2.30E-04	
262.1	1.61E-04	1.89E-04	370.1	2.32E-04	
264.1	1.62E-04	1.91E-04	372.1	2.33E-04	
266.1	1.63E-04	1.93E-04	374.1	2.34E-04	
268.1	1.65E-04	1.95E-04	376.1	2.36E-04	
270.1	1.66E-04	1.96E-04	378.1	2.37E-04	
272.1	1.67E-04	1.98E-04	380.1	2.38E-04	
274.1	1.69E-04	2.00E-04	382.1	2.40E-04	
276.1	1.70E-04	2.01E-04	384.1	2.41E-04	
278.1	1.71E-04	2.03E-04	386.1	2.42E-04	
280.1	1.72E-04	2.05E-04	388.1	2.44E-04	
282.1	1.74E-04	2.07E-04	390.1	2.45E-04	
284.1	1.75E-04	2.08E-04	392.1	2.46E-04	
286.1	1.76E-04	2.10E-04	394.1	2.48E-04	
288.1	1.78E-04	2.12E-04	396.1	2.49E-04	
290.1	1.79E-04	2.13E-04	398.1	2.50E-04	
292.1	1.80E-04	2.15E-04	400.1	2.52E-04	
294.1	1.82E-04	2.17E-04	402.1	2.53E-04	
296.1	1.83E-04	2.19E-04	404.1	2.54E-04	
298.1	1.84E-04	2.20E-04	406.1	2.56E-04	
300.1	1.86E-04	2.22E-04	408.1	2.57E-04	
302.1	1.87E-04	2.24E-04	410.1	2.58E-04	
304.1	1.88E-04	2.26E-04	412.1	2.60E-04	
306.1	1.89E-04	2.27E-04	414.1	2.61E-04	
308.1	1.91E-04	2.29E-04	416.1	2.62E-04	
310.1	1.92E-04	2.31E-04	418.1	2.64E-04	
312.1	1.93E-04	2.33E-04	420.1	2.65E-04	
314.1	1.95E-04	2.35E-04	422.1	2.66E-04	
316.1	1.96E-04	2.36E-04	424.1	2.68E-04	
318.1	1.97E-04	2.38E-04	426.1	2.69E-04	

Appendix 5: Different charges used for a constant plating time (chip 2)

	Plating 5	Plating 4		Plating 5	Plating 4		Plating 5	Plating 4
Time (s)	Charge (C)	Charge (C)	Time (s)	Charge (C)	Charge (C)	Time (s)	Charge (C)	Charge (C)
0.1	5.04E-08	1.31E-07	102.1	7.66E-05	5.54E-05	204.1	1.60E-04	1.10E-04
2.1	7.75E-07	1.42E-06	104.1	7.82E-05	5.65E-05	206.1	1.62E-04	1.11E-04
4.1	1.56E-06	2.59E-06	106.1	7.98E-05	5.76E-05	208.1	1.64E-04	1.12E-04
6.1	2.44E-06	3.73E-06	108.1	8.14E-05	5.86E-05	210.1	1.65E-04	1.13E-04
8.1	3.41E-06	4.85E-06	110.1	8.30E-05	5.97E-05	212.1	1.67E-04	1.14E-04
10.1	4.46E-06	5.95E-06	112.1	8.46E-05	6.07E-05	214.1	1.69E-04	1.15E-04
12.1	5.62E-06	7.04E-06	114.1	8.62E-05	6.18E-05	216.1	1.71E-04	1.16E-04
14.1	6.86E-06	8.13E-06	116.1	8.78E-05	6.29E-05	218.1	1.72E-04	1.17E-04
16.1	8.20E-06	9.22E-06	118.1	8.94E-05	6.39E-05	220.1	1.74E-04	1.18E-04
18.1	9.60E-06	1.03E-05	120.1	9.10E-05	6.50E-05	222.1	1.76E-04	1.19E-04
20.1	1.11E-05	1.14E-05	122.1	9.26E-05	6.61E-05	224.1	1.77E-04	1.21E-04
22.1	1.26E-05	1.25E-05	124.1	9.42E-05	6.71E-05	226.1	1.79E-04	1.22E-04
24.1	1.42E-05	1.35E-05	126.1	9.58E-05	6.82E-05	228.1	1.81E-04	1.23E-04
26.1	1.58E-05	1.46E-05	128.1	9.75E-05	6.93E-05	230.1	1.82E-04	1.24E-04
28.1	1.74E-05	1.57E-05	130.1	9.91E-05	7.03E-05	232.1	1.84E-04	1.25E-04
30.1	1.90E-05	1.68E-05	132.1	1.01E-04	7.14E-05	234.1	1.86E-04	1.26E-04
32.1	2.06E-05	1.78E-05	134.1	1.02E-04	7.25E-05	236.1	1.88E-04	1.27E-04
34.1	2.23E-05	1.89E-05	136.1	1.04E-04	7.35E-05	238.1	1.89E-04	1.28E-04
36.1	2.40E-05	2.00E-05	138.1	1.06E-04	7.46E-05	240.1	1.91E-04	1.29E-04
38.1	2.56E-05	2.11E-05	140.1	1.07E-04	7.56E-05	242.1	1.93E-04	1.30E-04
40.1	2.73E-05	2.21E-05	142.1	1.09E-04	7.67E-05	244.1	1.94E-04	1.31E-04
42.1	2.89E-05	2.32E-05	144.1	1.10E-04	7.78E-05	246.1	1.96E-04	1.32E-04
44.1	3.06E-05	2.43E-05	146.1	1.12E-04	7.88E-05	248.1	1.98E-04	1.33E-04
46.1	3.22E-05	2.54E-05	148.1	1.14E-04	7.99E-05	250.1	2.00E-04	1.34E-04
48.1	3.38E-05	2.64E-05	150.1	1.15E-04	8.10E-05	252.1	2.01E-04	1.36E-04
50.1	3.55E-05	2.75E-05	152.1	1.17E-04	8.20E-05	254.1	2.03E-04	1.37E-04
52.1	3.71E-05	2.86E-05	154.1	1.19E-04	8.31E-05	256.1	2.05E-04	1.38E-04
54.1	3.87E-05	2.97E-05	156.1	1.20E-04	8.42E-05	258.1	2.07E-04	1.39E-04
56.1	4.03E-05	3.07E-05	158.1	1.22E-04	8.52E-05	260.1	2.08E-04	1.40E-04
58.1	4.19E-05	3.18E-05	160.1	1.24E-04	8.63E-05	262.1	2.10E-04	1.41E-04
60.1	4.35E-05	3.29E-05	162.1	1.25E-04	8.74E-05	264.1	2.12E-04	1.42E-04
62.1	4.51E-05	3.39E-05	164.1	1.27E-04	8.84E-05	266.1	2.13E-04	1.43E-04
64.1	4.67E-05	3.50E-05	166.1	1.29E-04	8.95E-05	268.1	2.15E-04	1.44E-04
66.1	4.83E-05	3.61E-05	168.1	1.30E-04	9.06E-05	270.1	2.17E-04	1.45E-04
68.1	4.99E-05	3.72E-05	170.1	1.32E-04	9.16E-05	272.1	2.19E-04	1.46E-04
70.1	5.14E-05	3.83E-05	172.1	1.34E-04	9.27E-05	274.1	2.20E-04	1.47E-04
72.1	5.30E-05	3.93E-05	174.1	1.35E-04	9.38E-05	276.1	2.22E-04	1.48E-04
74.1	5.46E-05	4.04E-05	176.1	1.37E-04	9.48E-05	278.1	2.24E-04	1.50E-04
76.1	5.62E-05	4.15E-05	178.1	1.39E-04	9.59E-05	280.1	2.26E-04	1.51E-04
78.1	5.77E-05	4.26E-05	180.1	1.40E-04	9.70E-05	282.1	2.27E-04	1.52E-04
80.1	5.93E-05	4.36E-05	182.1	1.42E-04	9.80E-05	284.1	2.29E-04	1.53E-04
82.1	6.09E-05	4.47E-05	184.1	1.44E-04	9.91E-05	286.1	2.31E-04	1.54E-04
84.1	6.24E-05	4.58E-05	186.1	1.45E-04	1.00E-04	288.1	2.33E-04	1.55E-04
86.1	6.40E-05	4.69E-05	188.1	1.47E-04	1.01E-04	290.1	2.35E-04	1.56E-04
88.1	6.56E-05	4.79E-05	190.1	1.49E-04	1.02E-04	292.1	2.36E-04	1.57E-04
90.1	6.71E-05	4.90E-05	192.1	1.50E-04	1.03E-04	294.1	2.38E-04	1.58E-04
92.1	6.87E-05	5.01E-05	194.1	1.52E-04	1.04E-04	296.1	2.40E-04	1.59E-04
94.1	7.03E-05	5.12E-05	196.1	1.54E-04	1.05E-04	298.1	2.42E-04	1.60E-04
96.1	7.19E-05	5.22E-05	198.1	1.55E-04	1.07E-04	300.1	2.43E-04	1.61E-04
98.1	7.35E-05	5.33E-05	200.1	1.57E-04	1.08E-04			
100.1	7.50E-05	5.44E-05	202.1	1.59E-04	1.09E-04			

Appendix 6: Chronoamperometric response of the UMEA for a time frame of 1 s

t (s)	$1/t^{1/2}$	Current (A)	t (s)	$1/t^{1/2}$	Current (A)	t (s)	$1/t^{1/2}$	Current (A)	t (s)	$1/t^{1/2}$	Current (A)
0.001	31.6	-5.94E-06	0.235	2.1	-8.08E-07	0.505	1.4	-7.75E-07	0.775	1.1	-7.60E-07
0.002	22.4	-2.39E-06	0.240	2.0	-8.04E-07	0.510	1.4	-7.78E-07	0.780	1.1	-7.57E-07
0.003	18.3	-1.80E-06	0.245	2.0	-8.02E-07	0.515	1.4	-7.77E-07	0.785	1.1	-7.57E-07
0.004	15.8	-1.54E-06	0.250	2.0	-8.05E-07	0.520	1.4	-7.75E-07	0.790	1.1	-7.58E-07
0.005	14.1	-1.40E-06	0.255	2.0	-8.03E-07	0.525	1.4	-7.73E-07	0.795	1.1	-7.60E-07
0.006	12.9	-1.31E-06	0.260	2.0	-8.01E-07	0.530	1.4	-7.75E-07	0.800	1.1	-7.56E-07
0.007	12.0	-1.24E-06	0.265	1.9	-7.99E-07	0.535	1.4	-7.76E-07	0.805	1.1	-7.55E-07
0.008	11.2	-1.19E-06	0.270	1.9	-8.01E-07	0.540	1.4	-7.72E-07	0.810	1.1	-7.59E-07
0.009	10.5	-1.16E-06	0.275	1.9	-8.02E-07	0.545	1.4	-7.72E-07	0.815	1.1	-7.57E-07
0.010	10.0	-1.13E-06	0.280	1.9	-7.97E-07	0.550	1.3	-7.75E-07	0.820	1.1	-7.56E-07
0.015	8.2	-1.04E-06	0.285	1.9	-7.97E-07	0.555	1.3	-7.74E-07	0.825	1.1	-7.54E-07
0.020	7.1	-9.84E-07	0.290	1.9	-7.99E-07	0.560	1.3	-7.72E-07	0.830	1.1	-7.56E-07
0.025	6.3	-9.54E-07	0.295	1.8	-7.99E-07	0.565	1.3	-7.70E-07	0.835	1.1	-7.57E-07
0.030	5.8	-9.33E-07	0.300	1.8	-7.96E-07	0.570	1.3	-7.74E-07	0.840	1.1	-7.53E-07
0.035	5.3	-9.19E-07	0.305	1.8	-7.93E-07	0.575	1.3	-7.73E-07	0.845	1.1	-7.54E-07
0.040	5.0	-9.03E-07	0.310	1.8	-7.97E-07	0.580	1.3	-7.69E-07	0.850	1.1	-7.55E-07
0.045	4.7	-8.92E-07	0.315	1.8	-7.96E-07	0.585	1.3	-7.70E-07	0.855	1.1	-7.56E-07
0.050	4.5	-8.88E-07	0.320	1.8	-7.92E-07	0.590	1.3	-7.70E-07	0.860	1.1	-7.54E-07
0.055	4.3	-8.80E-07	0.325	1.8	-7.92E-07	0.595	1.3	-7.71E-07	0.865	1.1	-7.52E-07
0.060	4.1	-8.72E-07	0.330	1.7	-7.93E-07	0.600	1.3	-7.69E-07	0.870	1.1	-7.56E-07
0.065	3.9	-8.66E-07	0.335	1.7	-7.94E-07	0.605	1.3	-7.67E-07	0.875	1.1	-7.55E-07
0.070	3.8	-8.63E-07	0.340	1.7	-7.89E-07	0.610	1.3	-7.71E-07	0.880	1.1	-7.51E-07
0.075	3.7	-8.61E-07	0.345	1.7	-7.89E-07	0.615	1.3	-7.69E-07	0.885	1.1	-7.52E-07
0.080	3.5	-8.53E-07	0.350	1.7	-7.92E-07	0.620	1.3	-7.68E-07	0.890	1.1	-7.53E-07
0.085	3.4	-8.50E-07	0.355	1.7	-7.91E-07	0.625	1.3	-7.66E-07	0.895	1.1	-7.55E-07
0.090	3.3	-8.50E-07	0.360	1.7	-7.89E-07	0.630	1.3	-7.68E-07	0.900	1.1	-7.51E-07
0.095	3.2	-8.47E-07	0.365	1.7	-7.87E-07	0.635	1.3	-7.69E-07	0.905	1.1	-7.50E-07
0.100	3.2	-8.42E-07	0.370	1.6	-7.89E-07	0.640	1.2	-7.65E-07	0.910	1.0	-7.53E-07
0.105	3.1	-8.38E-07	0.375	1.6	-7.90E-07	0.645	1.2	-7.65E-07	0.915	1.0	-7.52E-07
0.110	3.0	-8.40E-07	0.380	1.6	-7.86E-07	0.650	1.2	-7.68E-07	0.920	1.0	-7.51E-07
0.115	2.9	-8.37E-07	0.385	1.6	-7.86E-07	0.655	1.2	-7.67E-07	0.925	1.0	-7.49E-07
0.120	2.9	-8.32E-07	0.390	1.6	-7.86E-07	0.660	1.2	-7.65E-07	0.930	1.0	-7.51E-07
0.125	2.8	-8.31E-07	0.395	1.6	-7.87E-07	0.665	1.2	-7.63E-07	0.935	1.0	-7.53E-07
0.130	2.8	-8.31E-07	0.400	1.6	-7.85E-07	0.670	1.2	-7.67E-07	0.940	1.0	-7.49E-07
0.135	2.7	-8.31E-07	0.405	1.6	-7.82E-07	0.675	1.2	-7.66E-07	0.945	1.0	-7.49E-07
0.140	2.7	-8.26E-07	0.410	1.6	-7.86E-07	0.680	1.2	-7.63E-07	0.950	1.0	-7.50E-07
0.145	2.6	-8.23E-07	0.415	1.6	-7.85E-07	0.685	1.2	-7.63E-07	0.955	1.0	-7.50E-07
0.150	2.6	-8.26E-07	0.420	1.5	-7.82E-07	0.690	1.2	-7.64E-07	0.960	1.0	-7.49E-07
0.155	2.5	-8.23E-07	0.425	1.5	-7.82E-07	0.695	1.2	-7.65E-07	0.965	1.0	-7.46E-07
0.160	2.5	-8.21E-07	0.430	1.5	-7.83E-07	0.700	1.2	-7.63E-07	0.970	1.0	-7.50E-07
0.165	2.5	-8.18E-07	0.435	1.5	-7.84E-07	0.705	1.2	-7.61E-07	0.975	1.0	-7.49E-07
0.170	2.4	-8.19E-07	0.440	1.5	-7.80E-07	0.710	1.2	-7.64E-07	0.980	1.0	-7.47E-07
0.175	2.4	-8.20E-07	0.445	1.5	-7.80E-07	0.715	1.2	-7.63E-07	0.985	1.0	-7.47E-07
0.180	2.4	-8.14E-07	0.450	1.5	-7.82E-07	0.720	1.2	-7.62E-07	0.990	1.0	-7.48E-07
0.185	2.3	-8.14E-07	0.455	1.5	-7.81E-07	0.725	1.2	-7.60E-07	0.995	1.0	-7.50E-07
0.190	2.3	-8.16E-07	0.460	1.5	-7.80E-07	0.730	1.2	-7.61E-07	1.000	1.0	-7.46E-07
0.195	2.3	-8.15E-07	0.465	1.5	-7.78E-07	0.735	1.2	-7.63E-07			
0.200	2.2	-8.11E-07	0.470	1.5	-7.81E-07	0.740	1.2	-7.59E-07			
0.205	2.2	-8.09E-07	0.475	1.5	-7.81E-07	0.745	1.2	-7.59E-07			
0.210	2.2	-8.12E-07	0.480	1.4	-7.77E-07	0.750	1.2	-7.61E-07			
0.215	2.2	-8.10E-07	0.485	1.4	-7.77E-07	0.755	1.2	-7.61E-07			
0.220	2.1	-8.06E-07	0.490	1.4	-7.78E-07	0.760	1.1	-7.60E-07			
0.225	2.1	-8.06E-07	0.495	1.4	-7.79E-07	0.765	1.1	-7.57E-07			
0.230	2.1	-8.07E-07	0.500	1.4	-7.77E-07	0.770	1.1	-7.61E-07			

Appendix 7: Chronoamperometric response of the UMEA for a time frame of 1000 s

t (s)	$1/t^{1/2}$	Current (A)	t (s)	$1/t^{1/2}$	Current (A)	t (s)	$1/t^{1/2}$	Current (A)	t (s)	$1/t^{1/2}$	Current (A)
1	1.00	-7.80E-07	260	0.06	-5.58E-07	525	0.04	-5.24E-07	785	0.04	-4.98E-07
5	0.45	-6.82E-07	265	0.06	-5.57E-07	530	0.04	-5.24E-07	790	0.04	-4.97E-07
10	0.32	-6.40E-07	275	0.06	-5.55E-07	535	0.04	-5.22E-07	795	0.04	-4.97E-07
15	0.26	-6.19E-07	280	0.06	-5.55E-07	540	0.04	-5.21E-07	800	0.04	-4.98E-07
20	0.22	-6.07E-07	285	0.06	-5.55E-07	270	0.06	-5.56E-07	805	0.04	-4.98E-07
25	0.20	-6.00E-07	290	0.06	-5.55E-07	550	0.04	-5.22E-07	810	0.04	-4.97E-07
30	0.18	-5.96E-07	295	0.06	-5.54E-07	555	0.04	-5.22E-07	815	0.04	-4.97E-07
35	0.17	-5.93E-07	300	0.06	-5.53E-07	560	0.04	-5.21E-07	820	0.03	-4.96E-07
40	0.16	-5.91E-07	305	0.06	-5.52E-07	565	0.04	-5.19E-07	545	0.04	-5.22E-07
45	0.15	-5.91E-07	310	0.06	-5.50E-07	570	0.04	-5.18E-07	825	0.03	-4.95E-07
50	0.14	-5.90E-07	315	0.06	-5.49E-07	575	0.04	-5.18E-07	830	0.03	-4.96E-07
55	0.13	-5.88E-07	320	0.06	-5.48E-07	580	0.04	-5.18E-07	835	0.03	-4.96E-07
60	0.13	-5.87E-07	325	0.06	-5.48E-07	585	0.04	-5.19E-07	840	0.03	-4.96E-07
65	0.12	-5.87E-07	330	0.06	-5.47E-07	590	0.04	-5.19E-07	845	0.03	-4.95E-07
70	0.12	-5.86E-07	335	0.05	-5.47E-07	595	0.04	-5.19E-07	850	0.03	-4.94E-07
75	0.12	-5.86E-07	340	0.05	-5.47E-07	600	0.04	-5.18E-07	855	0.03	-4.94E-07
80	0.11	-5.85E-07	345	0.05	-5.46E-07	605	0.04	-5.17E-07	860	0.03	-4.95E-07
85	0.11	-5.84E-07	350	0.05	-5.45E-07	610	0.04	-5.17E-07	865	0.03	-4.95E-07
90	0.11	-5.84E-07	355	0.05	-5.44E-07	615	0.04	-5.15E-07	870	0.03	-4.94E-07
95	0.10	-5.83E-07	360	0.05	-5.42E-07	620	0.04	-5.14E-07	875	0.03	-4.93E-07
100	0.10	-5.83E-07	365	0.05	-5.40E-07	625	0.04	-5.14E-07	880	0.03	-4.92E-07
105	0.10	-5.83E-07	370	0.05	-5.39E-07	630	0.04	-5.14E-07	885	0.03	-4.93E-07
110	0.10	-5.81E-07	375	0.05	-5.39E-07	635	0.04	-5.14E-07	890	0.03	-4.92E-07
115	0.09	-5.80E-07	380	0.05	-5.40E-07	640	0.04	-5.14E-07	895	0.03	-4.93E-07
120	0.09	-5.80E-07	385	0.05	-5.40E-07	645	0.04	-5.14E-07	900	0.03	-4.93E-07
125	0.09	-5.79E-07	390	0.05	-5.39E-07	650	0.04	-5.11E-07	905	0.03	-4.92E-07
130	0.09	-5.76E-07	395	0.05	-5.38E-07	655	0.04	-5.11E-07	910	0.03	-4.91E-07
135	0.09	-5.76E-07	400	0.05	-5.37E-07	660	0.04	-5.10E-07	915	0.03	-4.90E-07
140	0.08	-5.74E-07	405	0.05	-5.35E-07	665	0.04	-5.10E-07	920	0.03	-4.90E-07
145	0.08	-5.74E-07	410	0.05	-5.35E-07	670	0.04	-5.11E-07	925	0.03	-4.90E-07
150	0.08	-5.73E-07	415	0.05	-5.35E-07	675	0.04	-5.10E-07	930	0.03	-4.90E-07
155	0.08	-5.72E-07	420	0.05	-5.36E-07	680	0.04	-5.10E-07	935	0.03	-4.91E-07
160	0.08	-5.72E-07	425	0.05	-5.35E-07	685	0.04	-5.08E-07	940	0.03	-4.91E-07
165	0.08	-5.72E-07	430	0.05	-5.34E-07	690	0.04	-5.07E-07	945	0.03	-4.91E-07
170	0.08	-5.71E-07	435	0.05	-5.33E-07	695	0.04	-5.07E-07	950	0.03	-4.90E-07
175	0.08	-5.70E-07	440	0.05	-5.32E-07	700	0.04	-5.08E-07	955	0.03	-4.90E-07
180	0.07	-5.68E-07	445	0.05	-5.31E-07	705	0.04	-5.08E-07	960	0.03	-4.88E-07
185	0.07	-5.68E-07	450	0.05	-5.30E-07	710	0.04	-5.07E-07	965	0.03	-4.88E-07
190	0.07	-5.68E-07	455	0.05	-5.29E-07	715	0.04	-5.05E-07	970	0.03	-4.89E-07
195	0.07	-5.68E-07	460	0.05	-5.29E-07	720	0.04	-5.04E-07	975	0.03	-4.89E-07
200	0.07	-5.67E-07	465	0.05	-5.28E-07	725	0.04	-5.02E-07	980	0.03	-4.89E-07
205	0.07	-5.66E-07	470	0.05	-5.28E-07	730	0.04	-5.03E-07	985	0.03	-4.89E-07
210	0.07	-5.65E-07	475	0.05	-5.28E-07	735	0.04	-5.03E-07	990	0.03	-4.88E-07
215	0.07	-5.65E-07	480	0.05	-5.28E-07	740	0.04	-5.02E-07	995	0.03	-4.87E-07
220	0.07	-5.64E-07	485	0.05	-5.28E-07	745	0.04	-5.01E-07	1000	0.03	-4.86E-07
225	0.07	-5.63E-07	490	0.05	-5.27E-07	750	0.04	-4.99E-07			
230	0.07	-5.62E-07	495	0.04	-5.26E-07	755	0.04	-4.99E-07			
235	0.07	-5.61E-07	500	0.04	-5.25E-07	760	0.04	-4.99E-07			
240	0.06	-5.61E-07	505	0.04	-5.24E-07	765	0.04	-5.00E-07			
245	0.06	-5.61E-07	510	0.04	-5.24E-07	770	0.04	-5.00E-07			
250	0.06	-5.60E-07	515	0.04	-5.24E-07	775	0.04	-5.00E-07			
255	0.06	-5.59E-07	520	0.04	-5.25E-07	780	0.04	-5.00E-07			

Appendix 8: Comparison of the experimental currents with the empirical values (Shoup & Szabo 1984)

t	r	d	m	n	F	D	C	τ	θ	$f(\tau, \theta)$	I_{inlaid}	$I_{recessed}$	$I_{experiment}$
(s)	(m)	(m)			(c/mol)	(m^2/s)	(mol/m^3)				(A)	(A)	(A)
1	9.0E-07	2.5E-05	1012	1	9.6485E+04	7.84E-10	6	3871.6	0.9948	0.9542	1.58E-06	1.23E-06	7.17E-07
2	9.0E-07	2.5E-05	1012	1	9.6485E+04	7.84E-10	6	7743.2	0.9948	0.8725	1.44E-06	1.12E-06	6.80E-07
4	9.0E-07	2.5E-05	1012	1	9.6485E+04	7.84E-10	6	15486.4	0.9948	0.7592	1.26E-06	9.78E-07	6.43E-07
6	9.0E-07	2.5E-05	1012	1	9.6485E+04	7.84E-10	6	23229.6	0.9948	0.6844	1.13E-06	8.82E-07	6.23E-07
8	9.0E-07	2.5E-05	1012	1	9.6485E+04	7.84E-10	6	30972.8	0.9948	0.6301	1.04E-06	8.12E-07	6.08E-07
10	9.0E-07	2.5E-05	1012	1	9.6485E+04	7.84E-10	6	38716.0	0.9948	0.5881	9.72E-07	7.58E-07	5.98E-07
12	9.0E-07	2.5E-05	1012	1	9.6485E+04	7.84E-10	6	46459.3	0.9948	0.5543	9.17E-07	7.14E-07	5.89E-07
14	9.0E-07	2.5E-05	1012	1	9.6485E+04	7.84E-10	6	54202.5	0.9948	0.5262	8.70E-07	6.78E-07	5.83E-07
16	9.0E-07	2.5E-05	1012	1	9.6485E+04	7.84E-10	6	61945.7	0.9948	0.5024	8.31E-07	6.47E-07	5.78E-07
18	9.0E-07	2.5E-05	1012	1	9.6485E+04	7.84E-10	6	69688.9	0.9948	0.4818	7.97E-07	6.21E-07	5.74E-07
20	9.0E-07	2.5E-05	1012	1	9.6485E+04	7.84E-10	6	77432.1	0.9948	0.4637	7.67E-07	5.98E-07	5.71E-07
22	9.0E-07	2.5E-05	1012	1	9.6485E+04	7.84E-10	6	85175.3	0.9948	0.4477	7.40E-07	5.77E-07	5.68E-07
24	9.0E-07	2.5E-05	1012	1	9.6485E+04	7.84E-10	6	92918.5	0.9948	0.4334	7.17E-07	5.59E-07	5.66E-07
26	9.0E-07	2.5E-05	1012	1	9.6485E+04	7.84E-10	6	100661.7	0.9948	0.4204	6.95E-07	5.42E-07	5.64E-07
28	9.0E-07	2.5E-05	1012	1	9.6485E+04	7.84E-10	6	108404.9	0.9948	0.4087	6.76E-07	5.27E-07	5.63E-07
30	9.0E-07	2.5E-05	1012	1	9.6485E+04	7.84E-10	6	116148.1	0.9948	0.3979	6.58E-07	5.13E-07	5.62E-07
32	9.0E-07	2.5E-05	1012	1	9.6485E+04	7.84E-10	6	123891.4	0.9948	0.3880	6.42E-07	5.00E-07	5.61E-07
34	9.0E-07	2.5E-05	1012	1	9.6485E+04	7.84E-10	6	131634.6	0.9948	0.3789	6.26E-07	4.88E-07	5.59E-07
36	9.0E-07	2.5E-05	1012	1	9.6485E+04	7.84E-10	6	139377.8	0.9948	0.3704	6.12E-07	4.77E-07	5.58E-07
38	9.0E-07	2.5E-05	1012	1	9.6485E+04	7.84E-10	6	147121.0	0.9948	0.3625	5.99E-07	4.67E-07	5.58E-07
40	9.0E-07	2.5E-05	1012	1	9.6485E+04	7.84E-10	6	154864.2	0.9948	0.3551	5.87E-07	4.58E-07	5.57E-07
42	9.0E-07	2.5E-05	1012	1	9.6485E+04	7.84E-10	6	162607.4	0.9948	0.3482	5.76E-07	4.49E-07	5.57E-07
44	9.0E-07	2.5E-05	1012	1	9.6485E+04	7.84E-10	6	170350.6	0.9948	0.3417	5.65E-07	4.40E-07	5.56E-07
46	9.0E-07	2.5E-05	1012	1	9.6485E+04	7.84E-10	6	178093.8	0.9948	0.3355	5.55E-07	4.32E-07	5.56E-07
48	9.0E-07	2.5E-05	1012	1	9.6485E+04	7.84E-10	6	185837.0	0.9948	0.3297	5.45E-07	4.25E-07	5.56E-07
50	9.0E-07	2.5E-05	1012	1	9.6485E+04	7.84E-10	6	193580.2	0.9948	0.3243	5.36E-07	4.18E-07	5.55E-07
52	9.0E-07	2.5E-05	1012	1	9.6485E+04	7.84E-10	6	201323.5	0.9948	0.3190	5.28E-07	4.11E-07	5.55E-07
54	9.0E-07	2.5E-05	1012	1	9.6485E+04	7.84E-10	6	209066.7	0.9948	0.3141	5.19E-07	4.05E-07	5.55E-07
56	9.0E-07	2.5E-05	1012	1	9.6485E+04	7.84E-10	6	216809.9	0.9948	0.3094	5.12E-07	3.99E-07	5.56E-07
58	9.0E-07	2.5E-05	1012	1	9.6485E+04	7.84E-10	6	224553.1	0.9948	0.3049	5.04E-07	3.93E-07	5.56E-07
60	9.0E-07	2.5E-05	1012	1	9.6485E+04	7.84E-10	6	232296.3	0.9948	0.3006	4.97E-07	3.87E-07	5.56E-07
62	9.0E-07	2.5E-05	1012	1	9.6485E+04	7.84E-10	6	240039.5	0.9948	0.2965	4.90E-07	3.82E-07	5.55E-07
64	9.0E-07	2.5E-05	1012	1	9.6485E+04	7.84E-10	6	247782.7	0.9948	0.2925	4.84E-07	3.77E-07	5.55E-07
66	9.0E-07	2.5E-05	1012	1	9.6485E+04	7.84E-10	6	255525.9	0.9948	0.2888	4.77E-07	3.72E-07	5.55E-07
68	9.0E-07	2.5E-05	1012	1	9.6485E+04	7.84E-10	6	263269.1	0.9948	0.2851	4.71E-07	3.67E-07	5.55E-07
70	9.0E-07	2.5E-05	1012	1	9.6485E+04	7.84E-10	6	271012.3	0.9948	0.2816	4.66E-07	3.63E-07	5.55E-07
72	9.0E-07	2.5E-05	1012	1	9.6485E+04	7.84E-10	6	278755.6	0.9948	0.2783	4.60E-07	3.59E-07	5.55E-07
74	9.0E-07	2.5E-05	1012	1	9.6485E+04	7.84E-10	6	286498.8	0.9948	0.2751	4.55E-07	3.55E-07	5.56E-07
76	9.0E-07	2.5E-05	1012	1	9.6485E+04	7.84E-10	6	294242.0	0.9948	0.2719	4.50E-07	3.50E-07	5.56E-07
78	9.0E-07	2.5E-05	1012	1	9.6485E+04	7.84E-10	6	301985.2	0.9948	0.2689	4.45E-07	3.47E-07	5.55E-07
80	9.0E-07	2.5E-05	1012	1	9.6485E+04	7.84E-10	6	309728.4	0.9948	0.2660	4.40E-07	3.43E-07	5.55E-07
82	9.0E-07	2.5E-05	1012	1	9.6485E+04	7.84E-10	6	317471.6	0.9948	0.2632	4.35E-07	3.39E-07	5.55E-07
84	9.0E-07	2.5E-05	1012	1	9.6485E+04	7.84E-10	6	325214.8	0.9948	0.2605	4.31E-07	3.36E-07	5.56E-07
86	9.0E-07	2.5E-05	1012	1	9.6485E+04	7.84E-10	6	332958.0	0.9948	0.2579	4.26E-07	3.32E-07	5.56E-07
88	9.0E-07	2.5E-05	1012	1	9.6485E+04	7.84E-10	6	340701.2	0.9948	0.2553	4.22E-07	3.29E-07	5.57E-07
90	9.0E-07	2.5E-05	1012	1	9.6485E+04	7.84E-10	6	348444.4	0.9948	0.2528	4.18E-07	3.26E-07	5.57E-07
92	9.0E-07	2.5E-05	1012	1	9.6485E+04	7.84E-10	6	356187.7	0.9948	0.2504	4.14E-07	3.23E-07	5.56E-07
94	9.0E-07	2.5E-05	1012	1	9.6485E+04	7.84E-10	6	363930.9	0.9948	0.2481	4.10E-07	3.20E-07	5.56E-07
96	9.0E-07	2.5E-05	1012	1	9.6485E+04	7.84E-10	6	371674.1	0.9948	0.2458	4.07E-07	3.17E-07	5.57E-07
98	9.0E-07	2.5E-05	1012	1	9.6485E+04	7.84E-10	6	379417.3	0.9948	0.2436	4.03E-07	3.14E-07	5.57E-07
100	9.0E-07	2.5E-05	1012	1	9.6485E+04	7.84E-10	6	387160.5	0.9948	0.2415	3.99E-07	3.11E-07	5.57E-07
102	9.0E-07	2.5E-05	1012	1	9.6485E+04	7.84E-10	6	394903.7	0.9948	0.2394	3.96E-07	3.09E-07	5.57E-07
104	9.0E-07	2.5E-05	1012	1	9.6485E+04	7.84E-10	6	402646.9	0.9948	0.2374	3.93E-07	3.06E-07	5.57E-07
106	9.0E-07	2.5E-05	1012	1	9.6485E+04	7.84E-10	6	410390.1	0.9948	0.2354	3.89E-07	3.03E-07	5.57E-07
108	9.0E-07	2.5E-05	1012	1	9.6485E+04	7.84E-10	6	418133.3	0.9948	0.2335	3.86E-07	3.01E-07	5.57E-07
110	9.0E-07	2.5E-05	1012	1	9.6485E+04	7.84E-10	6	425876.5	0.9948	0.2316	3.83E-07	2.99E-07	5.58E-07
112	9.0E-07	2.5E-05	1012	1	9.6485E+04	7.84E-10	6	433619.8	0.9948	0.2298	3.80E-07	2.96E-07	5.57E-07
114	9.0E-07	2.5E-05	1012	1	9.6485E+04	7.84E-10	6	441363.0	0.9948	0.2280	3.77E-07	2.94E-07	5.57E-07

Continue

t	r	d	m	n	F	D	C	τ	θ	$f(\tau, \theta)$	I_{inlaid}	$I_{recessed}$	$I_{experiment}$
(s)	(m)	(m)			(c/mol)	(m^2/s)	(mol/m^3)				(A)	(A)	(A)
116	9.0E-07	2.5E-05	1012	1	9.6485E+04	7.84E-10	6	449106.2	0.9948	0.2263	3.74E-07	2.92E-07	5.56E-07
118	9.0E-07	2.5E-05	1012	1	9.6485E+04	7.84E-10	6	456849.4	0.9948	0.2246	3.71E-07	2.89E-07	5.56E-07
120	9.0E-07	2.5E-05	1012	1	9.6485E+04	7.84E-10	6	464592.6	0.9948	0.2229	3.69E-07	2.87E-07	5.55E-07
122	9.0E-07	2.5E-05	1012	1	9.6485E+04	7.84E-10	6	472335.8	0.9948	0.2213	3.66E-07	2.85E-07	5.55E-07
124	9.0E-07	2.5E-05	1012	1	9.6485E+04	7.84E-10	6	480079.0	0.9948	0.2197	3.63E-07	2.83E-07	5.55E-07
126	9.0E-07	2.5E-05	1012	1	9.6485E+04	7.84E-10	6	487822.2	0.9948	0.2181	3.61E-07	2.81E-07	5.55E-07
128	9.0E-07	2.5E-05	1012	1	9.6485E+04	7.84E-10	6	495565.4	0.9948	0.2166	3.58E-07	2.79E-07	5.54E-07
130	9.0E-07	2.5E-05	1012	1	9.6485E+04	7.84E-10	6	503308.6	0.9948	0.2151	3.56E-07	2.77E-07	5.54E-07
132	9.0E-07	2.5E-05	1012	1	9.6485E+04	7.84E-10	6	511051.9	0.9948	0.2137	3.53E-07	2.75E-07	5.51E-07
134	9.0E-07	2.5E-05	1012	1	9.6485E+04	7.84E-10	6	518795.1	0.9948	0.2123	3.51E-07	2.74E-07	5.51E-07
136	9.0E-07	2.5E-05	1012	1	9.6485E+04	7.84E-10	6	526538.3	0.9948	0.2109	3.49E-07	2.72E-07	5.51E-07
138	9.0E-07	2.5E-05	1012	1	9.6485E+04	7.84E-10	6	534281.5	0.9948	0.2095	3.46E-07	2.70E-07	5.51E-07
140	9.0E-07	2.5E-05	1012	1	9.6485E+04	7.84E-10	6	542024.7	0.9948	0.2082	3.44E-07	2.68E-07	5.52E-07
142	9.0E-07	2.5E-05	1012	1	9.6485E+04	7.84E-10	6	549767.9	0.9948	0.2068	3.42E-07	2.67E-07	5.52E-07
144	9.0E-07	2.5E-05	1012	1	9.6485E+04	7.84E-10	6	557511.1	0.9948	0.2056	3.40E-07	2.65E-07	5.52E-07
146	9.0E-07	2.5E-05	1012	1	9.6485E+04	7.84E-10	6	565254.3	0.9948	0.2043	3.38E-07	2.63E-07	5.52E-07
148	9.0E-07	2.5E-05	1012	1	9.6485E+04	7.84E-10	6	572997.5	0.9948	0.2031	3.36E-07	2.62E-07	5.52E-07
150	9.0E-07	2.5E-05	1012	1	9.6485E+04	7.84E-10	6	580740.7	0.9948	0.2018	3.34E-07	2.60E-07	5.52E-07
152	9.0E-07	2.5E-05	1012	1	9.6485E+04	7.84E-10	6	588484.0	0.9948	0.2006	3.32E-07	2.59E-07	5.52E-07
154	9.0E 07	2.5E-05	1012	1	9.6485E+04	7.84E-10	6	596227.2	0.9948	0.1995	3.30E-07	2.57E-07	5.52E-07
156	9.0E-07	2.5E-05	1012	1	9.6485E+04	7.84E-10	6	603970.4	0.9948	0.1983	3.28E-07	2.56E-07	5.52E-07
158	9.0E-07	2.5E-05	1012	1	9.6485E+04	7.84E-10	6	611713.6	0.9948	0.1972	3.26E-07	2.54E-07	5.52E-07
160	9.0E-07	2.5E-05	1012	1	9.6485E+04	7.84E-10	6	619456.8	0.9948	0.1961	3.24E-07	2.53E-07	5.52E-07
162	9.0E-07	2.5E-05	1012	1	9.6485E+04	7.84E-10	6	627200.0	0.9948	0.1950	3.22E-07	2.51E-07	5.52E-07
164	9.0E-07	2.5E-05	1012	1	9.6485E+04	7.84E-10	6	634943.2	0.9948	0.1939	3.21E-07	2.50E-07	5.52E-07
166	9.0E-07	2.5E-05	1012	1	9.6485E+04	7.84E-10	6	642686.4	0.9948	0.1928	3.19E-07	2.49E-07	5.52E-07
168	9.0E-07	2.5E-05	1012	1	9.6485E+04	7.84E-10	6	650429.6	0.9948	0.1918	3.17E-07	2.47E-07	5.51E-07
170	9.0E-07	2.5E-05	1012	1	9.6485E+04	7.84E-10	6	658172.8	0.9948	0.1908	3.15E-07	2.46E-07	5.51E-07
172	9.0E-07	2.5E-05	1012	1	9.6485E+04	7.84E-10	6	665916.0	0.9948	0.1898	3.14E-07	2.45E-07	5.51E-07
174	9.0E-07	2.5E-05	1012	1	9.6485E+04	7.84E-10	6	673659.3	0.9948	0.1888	3.12E-07	2.43E-07	5.50E-07
176	9.0E-07	2.5E-05	1012	1	9.6485E+04	7.84E-10	6	681402.5	0.9948	0.1878	3.11E-07	2.42E-07	5.50E-07
178	9.0E-07	2.5E-05	1012	1	9.6485E+04	7.84E-10	6	689145.7	0.9948	0.1869	3.09E-07	2.41E-07	5.51E-07
180	9.0E-07	2.5E-05	1012	1	9.6485E+04	7.84E-10	6	696888.9	0.9948	0.1859	3.07E-07	2.40E-07	5.51E-07
182	9.0E-07	2.5E-05	1012	1	9.6485E+04	7.84E-10	6	704632.1	0.9948	0.1850	3.06E-07	2.38E-07	5.51E-07
184	9.0E-07	2.5E-05	1012	1	9.6485E+04	7.84E-10	6	712375.3	0.9948	0.1841	3.04E-07	2.37E-07	5.50E-07
186	9.0E-07	2.5E-05	1012	1	9.6485E+04	7.84E-10	6	720118.5	0.9948	0.1832	3.03E-07	2.36E-07	5.50E-07
188	9.0E-07	2.5E-05	1012	1	9.6485E+04	7.84E-10	6	727861.7	0.9948	0.1823	3.01E-07	2.35E-07	5.50E-07
190	9.0E-07	2.5E-05	1012	1	9.6485E+04	7.84E-10	6	735604.9	0.9948	0.1814	3.00E-07	2.34E-07	5.50E-07
192	9.0E-07	2.5E-05	1012	1	9.6485E+04	7.84E-10	6	743348.1	0.9948	0.1806	2.99E-07	2.33E-07	5.49E-07
194	9.0E-07	2.5E-05	1012	1	9.6485E+04	7.84E-10	6	751091.4	0.9948	0.1797	2.97E-07	2.32E-07	5.48E-07
196	9.0E-07	2.5E-05	1012	1	9.6485E+04	7.84E-10	6	758834.6	0.9948	0.1789	2.96E-07	2.31E-07	5.48E-07
198	9.0E-07	2.5E-05	1012	1	9.6485E+04	7.84E-10	6	766577.8	0.9948	0.1781	2.94E-07	2.29E-07	5.47E-07
200	9.0E-07	2.5E-05	1012	1	9.6485E+04	7.84E-10	6	774321.0	0.9948	0.1773	2.93E-07	2.28E-07	5.47E-07
202	9.0E-07	2.5E-05	1012	1	9.6485E+04	7.84E-10	6	782064.2	0.9948	0.1765	2.92E-07	2.27E-07	5.47E-07
204	9.0E-07	2.5E-05	1012	1	9.6485E+04	7.84E-10	6	789807.4	0.9948	0.1757	2.90E-07	2.26E-07	5.47E-07
206	9.0E-07	2.5E-05	1012	1	9.6485E+04	7.84E-10	6	797550.6	0.9948	0.1749	2.89E-07	2.25E-07	5.46E-07
208	9.0E-07	2.5E-05	1012	1	9.6485E+04	7.84E-10	6	805293.8	0.9948	0.1741	2.88E-07	2.24E-07	5.46E-07
210	9.0E-07	2.5E-05	1012	1	9.6485E+04	7.84E-10	6	813037.0	0.9948	0.1734	2.87E-07	2.23E-07	5.46E-07
212	9.0E-07	2.5E-05	1012	1	9.6485E+04	7.84E-10	6	820780.2	0.9948	0.1726	2.85E-07	2.22E-07	5.45E-07
214	9.0E-07	2.5E-05	1012	1	9.6485E+04	7.84E-10	6	828523.5	0.9948	0.1719	2.84E-07	2.22E-07	5.45E-07
216	9.0E-07	2.5E-05	1012	1	9.6485E+04	7.84E-10	6	836266.7	0.9948	0.1711	2.83E-07	2.21E-07	5.44E-07
218	9.0E-07	2.5E-05	1012	1	9.6485E+04	7.84E-10	6	844009.9	0.9948	0.1704	2.82E-07	2.20E-07	5.44E-07
220	9.0E-07	2.5E-05	1012	1	9.6485E+04	7.84E-10	6	851753.1	0.9948	0.1697	2.81E-07	2.19E-07	5.44E-07
222	9.0E-07	2.5E-05	1012	1	9.6485E+04	7.84E-10	6	859496.3	0.9948	0.1690	2.79E-07	2.18E-07	5.44E-07
224	9.0E-07	2.5E-05	1012	1	9.6485E+04	7.84E-10	6	867239.5	0.9948	0.1683	2.78E-07	2.17E-07	5.44E-07
226	9.0E-07	2.5E-05	1012	1	9.6485E+04	7.84E-10	6	874982.7	0.9948	0.1676	2.77E-07	2.16E-07	5.44E-07
228	9.0E-07	2.5E-05	1012	1	9.6485E+04	7.84E-10	6	882725.9	0.9948	0.1670	2.76E-07	2.15E-07	5.44E-07
230	9.0E-07	2.5E-05	1012	1	9.6485E+04	7.84E-10	6	890469.1	0.9948	0.1663	2.75E-07	2.14E-07	5.44E-07
232	9.0E-07	2.5E-05	1012	1	9.6485E+04	7.84E-10	6	898212.3	0.9948	0.1656	2.74E-07	2.13E-07	5.44E-07
234	9.0E-07	2.5E-05	1012	1	9.6485E+04	7.84E-10	6	905955.6	0.9948	0.1650	2.73E-07	2.13E-07	5.44E-07

Continue

t	r	d	m	n	F	D	C	τ	θ	$f(\tau, \theta)$	I_{inlaid}	$I_{recessed}$	$I_{experiment}$
(s)	(m)	(m)			(c/mol)	(m^2/s)	(mol/m^3)				(A)	(A)	(A)
236	9.0E-07	2.5E-05	1012	1	9.6485E+04	7.84E-10	6	913698.8	0.9948	0.1644	2.72E-07	2.12E-07	5.44E-07
238	9.0E-07	2.5E-05	1012	1	9.6485E+04	7.84E-10	6	921442.0	0.9948	0.1637	2.71E-07	2.11E-07	5.43E-07
240	9.0E-07	2.5E-05	1012	1	9.6485E+04	7.84E-10	6	929185.2	0.9948	0.1631	2.70E-07	2.10E-07	5.42E-07
242	9.0E-07	2.5E-05	1012	1	9.6485E+04	7.84E-10	6	936928.4	0.9948	0.1625	2.69E-07	2.09E-07	5.43E-07
244	9.0E-07	2.5E-05	1012	1	9.6485E+04	7.84E-10	6	944671.6	0.9948	0.1619	2.68E-07	2.09E-07	5.42E-07
246	9.0E-07	2.5E-05	1012	1	9.6485E+04	7.84E-10	6	952414.8	0.9948	0.1613	2.67E-07	2.08E-07	5.42E-07
248	9.0E-07	2.5E-05	1012	1	9.6485E+04	7.84E-10	6	960158.0	0.9948	0.1607	2.66E-07	2.07E-07	5.42E-07
250	9.0E-07	2.5E-05	1012	1	9.6485E+04	7.84E-10	6	967901.2	0.9948	0.1601	2.65E-07	2.06E-07	5.42E-07
252	9.0E-07	2.5E-05	1012	1	9.6485E+04	7.84E-10	6	975644.4	0.9948	0.1595	2.64E-07	2.06E-07	5.41E-07
254	9.0E-07	2.5E-05	1012	1	9.6485E+04	7.84E-10	6	983387.7	0.9948	0.1589	2.63E-07	2.05E-07	5.41E-07
256	9.0E-07	2.5E-05	1012	1	9.6485E+04	7.84E-10	6	991130.9	0.9948	0.1583	2.62E-07	2.04E-07	5.41E-07
258	9.0E-07	2.5E-05	1012	1	9.6485E+04	7.84E-10	6	998874.1	0.9948	0.1578	2.61E-07	2.03E-07	5.41E-07
260	9.0E-07	2.5E-05	1012	1	9.6485E+04	7.84E-10	6	1006617.3	0.9948	0.1572	2.60E-07	2.03E-07	5.40E-07
262	9.0E-07	2.5E-05	1012	1	9.6485E+04	7.84E-10	6	1014360.5	0.9948	0.1566	2.59E-07	2.02E-07	5.40E-07
264	9.0E-07	2.5E-05	1012	1	9.6485E+04	7.84E-10	6	1022103.7	0.9948	0.1561	2.58E-07	2.01E-07	5.40E-07
266	9.0E-07	2.5E-05	1012	1	9.6485E+04	7.84E-10	6	1029846.9	0.9948	0.1556	2.57E-07	2.00E-07	5.40E-07
268	9.0E-07	2.5E-05	1012	1	9.6485E+04	7.84E-10	6	1037590.1	0.9948	0.1550	2.56E-07	2.00E-07	5.41E-07
270	9.0E-07	2.5E-05	1012	1	9.6485E+04	7.84E-10	6	1045333.3	0.9948	0.1545	2.55E-07	1.99E-07	5.41E-07
272	9.0E-07	2.5E-05	1012	1	9.6485E+04	7.84E-10	6	1053076.5	0.9948	0.1540	2.55E-07	1.98E-07	5.41E-07
274	9.0E-07	2.5E-05	1012	1	9.6485E+04	7.84E-10	6	1060819.8	0.9948	0.1534	2.54E-07	1.98E-07	5.41E-07
276	9.0E-07	2.5E-05	1012	1	9.6485E+04	7.84E-10	6	1068563.0	0.9948	0.1529	2.53E-07	1.97E-07	5.40E-07
278	9.0E-07	2.5E-05	1012	1	9.6485E+04	7.84E-10	6	1076306.2	0.9948	0.1524	2.52E-07	1.96E-07	5.41E-07
280	9.0E-07	2.5E-05	1012	1	9.6485E+04	7.84E-10	6	1084049.4	0.9948	0.1519	2.51E-07	1.96E-07	5.41E-07
282	9.0E-07	2.5E-05	1012	1	9.6485E+04	7.84E-10	6	1091792.6	0.9948	0.1514	2.50E-07	1.95E-07	5.41E-07
284	9.0E-07	2.5E-05	1012	1	9.6485E+04	7.84E-10	6	1099535.8	0.9948	0.1509	2.50E-07	1.95E-07	5.41E-07
286	9.0E-07	2.5E-05	1012	1	9.6485E+04	7.84E-10	6	1107279.0	0.9948	0.1504	2.49E-07	1.94E-07	5.42E-07
288	9.0E-07	2.5E-05	1012	1	9.6485E+04	7.84E-10	6	1115022.2	0.9948	0.1500	2.48E-07	1.93E-07	5.41E-07
290	9.0E-07	2.5E-05	1012	1	9.6485E+04	7.84E-10	6	1122765.4	0.9948	0.1495	2.47E-07	1.93E-07	5.41E-07
292	9.0E-07	2.5E-05	1012	1	9.6485E+04	7.84E-10	6	1130508.6	0.9948	0.1490	2.46E-07	1.92E-07	5.41E-07
294	9.0E-07	2.5E-05	1012	1	9.6485E+04	7.84E-10	6	1138251.9	0.9948	0.1485	2.46E-07	1.91E-07	5.39E-07
296	9.0E-07	2.5E-05	1012	1	9.6485E+04	7.84E-10	6	1145995.1	0.9948	0.1481	2.45E-07	1.91E-07	5.38E-07
298	9.0E-07	2.5E-05	1012	1	9.6485E+04	7.84E-10	6	1153738.3	0.9948	0.1476	2.44E-07	1.90E-07	5.38E-07
300	9.0E-07	2.5E-05	1012	1	9.6485E+04	7.84E-10	6	1161481.5	0.9948	0.1472	2.43E-07	1.90E-07	5.37E-07
302	9.0E-07	2.5E-05	1012	1	9.6485E+04	7.84E-10	6	1169224.7	0.9948	0.1467	2.43E-07	1.89E-07	5.37E-07
304	9.0E-07	2.5E-05	1012	1	9.6485E+04	7.84E-10	6	1176967.9	0.9948	0.1463	2.42E-07	1.88E-07	5.38E-07
306	9.0E-07	2.5E-05	1012	1	9.6485E+04	7.84E-10	6	1184711.1	0.9948	0.1458	2.41E-07	1.88E-07	5.38E-07
308	9.0E-07	2.5E-05	1012	1	9.6485E+04	7.84E-10	6	1192454.3	0.9948	0.1454	2.40E-07	1.87E-07	5.38E-07
310	9.0E-07	2.5E-05	1012	1	9.6485E+04	7.84E-10	6	1200197.5	0.9948	0.1449	2.40E-07	1.87E-07	5.37E-07
312	9.0E-07	2.5E-05	1012	1	9.6485E+04	7.84E-10	6	1207940.7	0.9948	0.1445	2.39E-07	1.86E-07	5.37E-07
314	9.0E-07	2.5E-05	1012	1	9.6485E+04	7.84E-10	6	1215684.0	0.9948	0.1441	2.38E-07	1.86E-07	5.37E-07
316	9.0E-07	2.5E-05	1012	1	9.6485E+04	7.84E-10	6	1223427.2	0.9948	0.1437	2.38E-07	1.85E-07	5.37E-07
318	9.0E-07	2.5E-05	1012	1	9.6485E+04	7.84E-10	6	1231170.4	0.9948	0.1432	2.37E-07	1.85E-07	5.36E-07
320	9.0E-07	2.5E-05	1012	1	9.6485E+04	7.84E-10	6	1238913.6	0.9948	0.1428	2.36E-07	1.84E-07	5.37E-07
322	9.0E-07	2.5E-05	1012	1	9.6485E+04	7.84E-10	6	1246656.8	0.9948	0.1424	2.35E-07	1.84E-07	5.36E-07
324	9.0E-07	2.5E-05	1012	1	9.6485E+04	7.84E-10	6	1254400.0	0.9948	0.1420	2.35E-07	1.83E-07	5.36E-07
326	9.0E-07	2.5E-05	1012	1	9.6485E+04	7.84E-10	6	1262143.2	0.9948	0.1416	2.34E-07	1.82E-07	5.35E-07
328	9.0E-07	2.5E-05	1012	1	9.6485E+04	7.84E-10	6	1269886.4	0.9948	0.1412	2.33E-07	1.82E-07	5.36E-07
330	9.0E-07	2.5E-05	1012	1	9.6485E+04	7.84E-10	6	1277629.6	0.9948	0.1408	2.33E-07	1.81E-07	5.35E-07
332	9.0E-07	2.5E-05	1012	1	9.6485E+04	7.84E-10	6	1285372.8	0.9948	0.1404	2.32E-07	1.81E-07	5.35E-07
334	9.0E-07	2.5E-05	1012	1	9.6485E+04	7.84E-10	6	1293116.0	0.9948	0.1400	2.32E-07	1.80E-07	5.35E-07
336	9.0E-07	2.5E-05	1012	1	9.6485E+04	7.84E-10	6	1300859.3	0.9948	0.1396	2.31E-07	1.80E-07	5.34E-07
338	9.0E-07	2.5E-05	1012	1	9.6485E+04	7.84E-10	6	1308602.5	0.9948	0.1392	2.30E-07	1.79E-07	5.35E-07
340	9.0E-07	2.5E-05	1012	1	9.6485E+04	7.84E-10	6	1316345.7	0.9948	0.1389	2.30E-07	1.79E-07	5.35E-07
342	9.0E-07	2.5E-05	1012	1	9.6485E+04	7.84E-10	6	1324088.9	0.9948	0.1385	2.29E-07	1.78E-07	5.36E-07
344	9.0E-07	2.5E-05	1012	1	9.6485E+04	7.84E-10	6	1331832.1	0.9948	0.1381	2.28E-07	1.78E-07	5.35E-07
346	9.0E-07	2.5E-05	1012	1	9.6485E+04	7.84E-10	6	1339575.3	0.9948	0.1377	2.28E-07	1.78E-07	5.35E-07
348	9.0E-07	2.5E-05	1012	1	9.6485E+04	7.84E-10	6	1347318.5	0.9948	0.1374	2.27E-07	1.77E-07	5.35E-07
350	9.0E-07	2.5E-05	1012	1	9.6485E+04	7.84E-10	6	1355061.7	0.9948	0.1370	2.27E-07	1.77E-07	5.35E-07
352	9.0E-07	2.5E-05	1012	1	9.6485E+04	7.84E-10	6	1362804.9	0.9948	0.1366	2.26E-07	1.76E-07	5.35E-07
354	9.0E-07	2.5E-05	1012	1	9.6485E+04	7.84E-10	6	1370548.1	0.9948	0.1363	2.25E-07	1.76E-07	5.35E-07

Continue

t	r	d	m	n	F	D	C	τ	θ	$f(\tau, \theta)$	I_{inlaid}	$I_{recessed}$	$I_{experiment}$
(s)	(m)	(m)			(c/mol)	(m^2/s)	(mol/m^3)				(A)	(A)	(A)
356	9.0E-07	2.5E-05	1012	1	9.6485E+04	7.84E-10	6	1378291.4	0.9948	0.1359	2.25E-07	1.75E-07	5.36E-07
358	9.0E-07	2.5E-05	1012	1	9.6485E+04	7.84E-10	6	1386034.6	0.9948	0.1356	2.24E-07	1.75E-07	5.36E-07
360	9.0E-07	2.5E-05	1012	1	9.6485E+04	7.84E-10	6	1393777.8	0.9948	0.1352	2.24E-07	1.74E-07	5.36E-07
362	9.0E-07	2.5E-05	1012	1	9.6485E+04	7.84E-10	6	1401521.0	0.9948	0.1349	2.23E-07	1.74E-07	5.36E-07
364	9.0E-07	2.5E-05	1012	1	9.6485E+04	7.84E-10	6	1409264.2	0.9948	0.1345	2.22E-07	1.73E-07	5.36E-07
366	9.0E-07	2.5E-05	1012	1	9.6485E+04	7.84E-10	6	1417007.4	0.9948	0.1342	2.22E-07	1.73E-07	5.36E-07
368	9.0E-07	2.5E-05	1012	1	9.6485E+04	7.84E-10	6	1424750.6	0.9948	0.1338	2.21E-07	1.72E-07	5.36E-07
370	9.0E-07	2.5E-05	1012	1	9.6485E+04	7.84E-10	6	1432493.8	0.9948	0.1335	2.21E-07	1.72E-07	5.36E-07
372	9.0E-07	2.5E-05	1012	1	9.6485E+04	7.84E-10	6	1440237.0	0.9948	0.1332	2.20E-07	1.72E-07	5.35E-07
374	9.0E-07	2.5E-05	1012	1	9.6485E+04	7.84E-10	6	1447980.2	0.9948	0.1328	2.20E-07	1.71E-07	5.35E-07
376	9.0E-07	2.5E-05	1012	1	9.6485E+04	7.84E-10	6	1455723.5	0.9948	0.1325	2.19E-07	1.71E-07	5.34E-07
378	9.0E-07	2.5E-05	1012	1	9.6485E+04	7.84E-10	6	1463466.7	0.9948	0.1322	2.19E-07	1.70E-07	5.33E-07
380	9.0E-07	2.5E-05	1012	1	9.6485E+04	7.84E-10	6	1471209.9	0.9948	0.1318	2.18E-07	1.70E-07	5.33E-07
382	9.0E-07	2.5E-05	1012	1	9.6485E+04	7.84E-10	6	1478953.1	0.9948	0.1315	2.17E-07	1.70E-07	5.33E-07
384	9.0E-07	2.5E-05	1012	1	9.6485E+04	7.84E-10	6	1486696.3	0.9948	0.1312	2.17E-07	1.69E-07	5.33E-07
386	9.0E-07	2.5E-05	1012	1	9.6485E+04	7.84E-10	6	1494439.5	0.9948	0.1309	2.16E-07	1.69E-07	5.33E-07
388	9.0E-07	2.5E-05	1012	1	9.6485E+04	7.84E-10	6	1502182.7	0.9948	0.1306	2.16E-07	1.68E-07	5.32E-07
390	9.0E-07	2.5E-05	1012	1	9.6485E+04	7.84E-10	6	1509925.9	0.9948	0.1302	2.15E-07	1.68E-07	5.31E-07
392	9.0E-07	2.5E-05	1012	1	9.6485E+04	7.84E-10	6	1517669.1	0.9948	0.1299	2.15E-07	1.67E-07	5.32E-07
394	9.0E-07	2.5E-05	1012	1	9.6485E+04	7.84E-10	6	1525412.3	0.9948	0.1296	2.14E-07	1.67E-07	5.32E-07
396	9.0E-07	2.5E-05	1012	1	9.6485E+04	7.84E-10	6	1533155.6	0.9948	0.1293	2.14E 07	1.67E-07	5.33E-07
398	9.0E-07	2.5E-05	1012	1	9.6485E+04	7.84E-10	6	1540898.8	0.9948	0.1290	2.13E-07	1.66E-07	5.33E-07
400	9.0E-07	2.5E-05	1012	1	9.6485E+04	7.84E-10	6	1548642.0	0.9948	0.1287	2.13E-07	1.66E-07	5.33E-07

In der Reihe "Karlsruher Geochemische Hefte" (ISSN 0943-8599) sind erschienen:

Band 1: U. Kramar (1993)
Methoden zur Interpretation von Daten der geochemischen Bachsedimentprospektion am Beispiel der Sierra de San Carlos/ Tamaulipas Mexiko

Band 2: Z. Berner (1993)
S-Isotopengeochemie in der KTB - Vorbohrung und Beziehungen zu den Spurenelementmustern der Pyrite

Band 3: J.-D. Eckhardt (1993)
Geochemische Untersuchungen an jungen Sedimenten von der Galapagos-Mikroplatte:
Hydrothermale und stratigraphisch signifikante Signale

Band 4: B. Bergfeldt (1994)
Lösungs- und Austauschprozesse in der ungesättigten Bodenwasserzone und Auswirkungen auf das Grundwasser

Band 5: M. Hodel (1994)
Untersuchungen zur Festlegung und Mobilisierung der Elemente As, Cd, Ni und Pb an ausgewählten Festphasen unter besonderer Berücksichtigung des Einflusses von Huminstoffen.

Band 6: T. Bergfeldt (1995)
Untersuchungen der Arsen- und Schwermetallmobilität in Bergbauhalden und kontaminierten Böden im Gebiet des Mittleren Schwarzwaldes.

Band 7: M. Manz (1995)
Umweltbelastungen durch Arsen und Schwermetalle in Böden, Halden, Pflanzen und Schlacken ehemaliger Bergbaugebiete des Mittleren und Südlichen Schwarzwaldes.

Band 8: J. Ritter (1995)
Genese der Mineralisation Herrmanngang im Albtalgranit (SE-Schwarzwald) und Wechselwirkungen mit dem Nebengestein.

Band 9: J. Castro (1995)
Umweltauswirkungen des Bergbaus im semiariden Gebiet von Santa Maria de la Paz, Mexiko.

Band 10: T. Rüde (1996)
Beiträge zur Geochemie des Arsens.

Band 11: J. Schäfer (1998)
Einträge und Kontaminationspfade Kfz-emittierter Platin-Gruppen-Elemente (PGE) in verschiedenen Umweltkompartimenten.

Band 12: M. A. Leosson (1999)
A Contribution to the Sulphur Isotope Geochemistry of the Upper Continental Crust: The KTB Main Hole - A Case Study

Band 13: B. A. Kappes (2000)
Mobilisierbarkeit von Schwermetallen und Arsen durch saure Grubenabwässer in Bergbaualtlasten der Ag-Pb-Zn-Lagerstätte in Wiesloch

Band 14: H. Philipp (2000)
The behaviour of platinum-group elements in petrogenetic processes:
A case study from the seaward-dipping reflector sequence (SDRS), Southeast Greenland volcanic rifted margin

Band 15: E. Walpersdorf (2000)
Nähr- und Spurenelementdynamik im Sediment/Wasser-Kontaktbereich nach einer Seekreideaufspülung - Pilotstudie Arendsee -

Band 16: R. H. Kölbl (2000)
Models of hydrothermal plumes by submarine diffuse venting in a coastal area: A case study for Milos, South Aegean Volcanic Arc, Greece

Band 17: U. Heiser (2000)
Calcium-rich Rhodochrosite layers in Sediments of the Gotland Deep, Baltic Sea, as Indicators for Seawater Inflow

In der Fortsetzungsreihe "Karlsruher Mineralogische und Geochemische Hefte" (ISSN 1618-2677) sind bisher erschienen:

Band 18: S. Norra (2001)
Umweltgeochemische Signale urbaner Systeme am Beispiel von Böden, Pflanzen, und Stäuben in Karlsruhe

Band 19: M. von Wagner, Alexander Salichow, Doris Stüben (2001)
Geochemische Reinigung kleiner Fließgewässer mit Mangankiesen, einem Abfallsprodukt aus Wasserwerken (GReiFMan)

Band 20: U. Berg (2002)
Die Kalzitapplikation als Restaurierungsmaßnahme für eutrophe Seen – ihre Optimierung und Bewertung

Band 21: Ch. Menzel (2002)
Bestimmung und Verteilung aquatischer PGE-Spezies in urbanen Systemen

Band 22: P. Graf (2002)
Meta-Kaolinit als latent-hydraulische Komponente in Kalkmörtel

Band 23: D. Buqezi-Ahmeti (2003)
Die Fluidgehalte der Mantel-Xenolithe des Alkaligesteins-Komplexes der Persani-Berge, Ostkarpaten, Rumänien: Untersuchungen an Fluid-Einschlüssen

Band 24: B. Scheibner (2003)
Das geochemische Verhalten der Platingruppenelemente bei der Entstehung und Differenzierung der Magmen der Kerguelen-Flutbasaltprovinz (Indischer Ozean)

Ab Band 25 erscheinen die Karlsruher Mineralogischen und Geochemischen Hefte im Karlsruher Universitätsverlag online unter der Internetadresse

http://www.uvka.de/univerlag/institut.php?fakultaet=1

Auf Wunsch sind beim Karlsruher Universitätsverlag gedruckte Exemplare erhältlich („print on demand").

Band 25: A. Stögbauer (2005)
Schwefelisotopenuntersuchungen in abwasserbelasteten Sedimenten Biogeochemische Umsetzungen und deren Auswirkung auf den Schwermetallhaushalt

Band 26: X. Xie (2005)
Assessment of an ultramicroelectrode array (UMEA) sensor for the determination of trace concentrations of heavy metals in water

Band 27: M. Harting (2005)
Zum Kreide/Tertiär-Übergang in NE-Mexiko: Geochemische Charakterisierung der Chicxulub-Impaktejekta

Band 28: M. Meudt (2005)
Geochemie der Foraminiferenschalen: Relevanz für die Rekonstruktion von Paläoumweltbedingungen

Printed in Dunstable, United Kingdom

91343138R00087